Norvald Kjerstad

ICE NAVIGATION

Norvald Kjerstad

ICE NAVIGATION

1st edition

tapir academic press

ISBN   978-82-519-2760-4

Graphic design and adaption: The Author
Cover Layout: The Author and Tapir Academic Press
Paper: Munken Polar 90 g
Printed and binded by: AIT Oslo AS
Translation to English: Anne Colling
Photo cover: Courtesy of TransAtlantic

This book has been published with founding from The University of Tromsø, NORWAY

Supplement to lecturers:
CD with all figures in PowerPoint format.
Contact publisher for distribution.

Tapir Academic Press publishes textbooks and academic literature for universities and university colleges, as well as for vocational and professional education. We also publish high quality literature of a more general nature. Our main product lines are:

- *Textbooks for higher education*
- *Research and reference literature*
- *Non-fiction*

We only use environmentally certified printing houses.

Tapir Academic Press
NO–7005 Trondheim, Norway
Tel.:   + 47 73 59 32 10
Email: post@tapirforlag.no
www.tapirforlag.no

Publishing Editor: lasse.postmyr@tapirforlag.no

Foreword

Within the professional areas of *Operation and management of ships, Sailing and manoeuvring* and *Navigation control*, which we at Ålesund University College now offer as a part of the subject *Navigation 3 with a navigation simulator,* obtaining suitable literature has been a problem for many years. I therefore started systematic work to develop a textbook which can be used in the subject. This Ice Navigation book is basically an englich version of part IV of that book which is entitled *Fremføring av skip med navigasjonskontroll, 2. Ed. (Sailing of ships with navigation control).* Emphasis is laid on exceeding the minimum requirements that are described in the STCW Convention as well as the IMO Guidelines for ships operating in Polar waters. The work with this part of the book really started in 1989 when I worked for Tromsø University College (now University of Tromsø) with a compendium in Arctic Operations.

The book contains a minimum of description of electronic navigation systems. Where it is relevant, reference is made to the book *Elektroniske og akustiske navigasjonssystemer, (Electronic and acoustic navigation systems),* which I have had published by Tapir akademisk forlag (4rd Ed. 2010) (http://butikk.tapirforlag.no/no/search/node/Kjerstad) .

For teachers or others who wish to use the book for instruction, the CD with all the updated figures can be obtained from the author (nk@hials.no) in PowerPoint format and in colour.

Preparation of a lot of the material is based on own experience and conversations with sailors on a range of different ships. This has been possible *inter alia* because some shipping companies have allowed me to join ships for longer or shorter periods as crew member, researcher or observer. Here I would especially mention the shipping companies Murmansk Shipping Company, Fednav, Ugland Shipping, Farstad Shipping, Norwegian Coast Guard Squadron North, and Wilhelmsen. Close contact with the Norwegian Coastal Administration, the Swedish Coastal Administration, Transport Canada, Rolls-Royce Marine and Ulstein Verft (Ulstein Yard) has also been very useful.

Ålesund University College, January 2011.

Norvald Kjerstad,
Professor, Nautical Science
(Professor-II, University of Tromsø)

Ålesund University College
6025 Ålesund
(Tel. +47 7016 1200, Email: nk@hials.no)

Contents

ICE NAVIGATION

1 Introduction and history

Introduction

Most of the present rules and regulations in relation to safety at sea originate in a way from problems connected to ice. As is known, the SOLAS Convention was introduced as a result of the *Titanic* accident in 1912. As is known, this loss occurred as the result of a collision with an iceberg. In the STCW Code, which describes the international minimum requirements for maritime education, it is stated that one *shall be able to operate ships in waters with ice*. In relation to the detailing describing other themes, this provides little guidance for schools which are to train navigators. Since in most countries there is little shipping which comes into contact with ice-covered waters, the schools have therefore not laid great emphasis on the challenges and hazards connected to sailing in ice-covered waters. For the same reason there has been very little literature available regarding ice navigation. Gradually, as shipping in arctic regions has long been in strong growth, many countries have spoken out for new and extensive requirements regarding competence of sailors and shipping companies which are to operate in these inhospitable regions. In the 1900s, therefore, the *Circumpolar Advisory Group on Ice Operations* (CAGIO) was formed. The Group consisted of representatives from all countries that bordered on ice-covered waters or had special interests in relation to Arctic / Antarctic shipping. CAGIO worked on two levels:

- Draw up international rules for operation of ships in Polar Regions. This included specific requirements for training of sailors.
- Draw up harmonised rules for ice-reinforcement of ships in all classification companies organised in IACS.

It was the intention that the Group's work should end in a Polar Code which could be ratified by IMO. However, it proved to be difficult to get a consensus for a new "Code", but ended instead as IMO guidelines (2002). However, the work with the marine engineering area was accepted by all classification companies, which means that ships that were built in the future with ice-reinforcement shall be based on a common understanding of strength calculations – harmonised IACS rules.

After this work, some institutes chose to offer courses in ice navigation in accordance with what is specified in the recommendations by IMO. In 2009, IMO decided that the requirement for ice competence should be heightened by implementing this in the B part of the STCW Code, as well as that the work with a Polar Code should be resumed. Det Norske Veritas and other classification societies has also directed greater focus on competence for operation of ships in areas where one can encounter ice. The subjects which are described in SeaSkills at DNV also have great similarities with what is outlined in the recommendations from IMO. On this basis we have chosen to dedicate this textbook to operation of ships in arctic regions. The extent is adapted to the recommendations made by IMO and will therefore go considerably further than that specified in STCW.

When one is to study and approach the problem area connected to sailing in ice-covered waters, it is both useful and interesting to learn about the historical aspects of such voyages. We therefore start with a short description of some of the most significant voyages and the motivation for them. There is much toil and tragedies behind the knowledge we have today regarding the Polar Regions. Nevertheless, in all our admiration of the old pioneers, we shall keep in mind that in the whole of the arctic area we are talking about, there have been people who have lived there. People have lived on the coast of Siberia, the coast of Canada and on Greenland for centuries, and have adapted to the harsh environment.

1.1 The time of the Pioneers

The Arctic

The urge to explore the unknown and perhaps develop new trade has for several centuries drawn explorers and traders to the arctic regions. The first of the arctic explorers we know of was the sailor, Pytheas, who probably reached Iceland or Northern Norway in 330 BC. The north-Norwegian Viking chieftain, Ottar, was also early. He travelled northeast in 890 to explore the extent of the country. It is known that he reached the White Sea. A few years later, in 982, Eirik the Red discovered Greeland, and settled on the southwest side of the new country. Eighteen years later, Leif Eriksson, son of Eirik the Red, sailed westwards and discovered *Vinland*. As a result of this, a settlement was built up on the northern tip of what today is Newfoundland. Based on the travels of the Greenlanders, a chart was made on which names such as Vinland, Markland and Helleland appear. This is a chart which is used on travels in the future up to the 1500s. During this period the climate in the Arctic

was relatively warm, and there is a lot that indicates that the ice conditions were relatively favourable.

Several hundred years later, at the end of the 16th century, the British and Dutch started looking at the possibility of finding a northern passage to the East. The driving power was to outstrip the Spanish and Portuguese who by that time controlled the sea traffic to the Far East. In order to understand the enormous effort, one must remember that neither the Suez nor Panama canals existed, and the known route to the East was extremely long.

One of the first expeditions was organised by the Portuguese and Danish states in 1476-77. The leader of this expedition was the Norwegian pilot Jon Skolp, and with him as Mate he probably had the not unknown Christopher Colombus. Colombus, who probably had Norwegian connections presumably received an understanding of America's extent on this trip, which in 1492 led him to the "discovery" of America. Skolp's expedition probably came to the Lancaster Sound at 73° north. In the period following, there are a series of new expeditions:

Sir Willoughby and Chancelor voyage in 1533 to find the Northeast Passage and in 1535 Cartier charts the coast around the St. Lawrence Gulf in Canada. In 1576 Frobisher searches for the Northwest Passage, and one after the other towards 1631 we find voyages taken by Davis, Button, Baffin and Foxe on the same errand.

Another relatively well documented was carried out by the norwegian Jens Munk. In 1618 he sails with two ships to find a northwesterly route to the East. Jens was at this time captain under king Christian IV in Denmark. The chart which was available, and which most probably was used on the trip is shown at Figure 1.1. The story of this particularly dramatic and rigorous voyage is one of the most exciting in Norwegian polar history. After having wintered in the Hudson Bay, Munk and two of his crew managed to sail back to Norway. The other 62 who had sailed out the year before were dead – most of them from scurvy.

In 1594-97, Wilhelm Barents voyaged to find the Northeast Passage (Figure 1.2). Bear Island and Svalbard were discovered on these trips. On this trip Barents also finds several traces of Russian hunters on Novaya Zemlya.

Figure 1.1
World chart which was available in Jens Munk's time. Note the passages in the North.

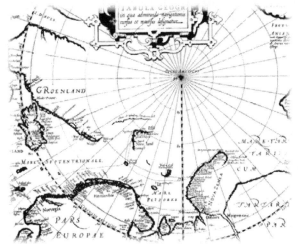

Figure 1.2
Chart of Novaya Zemlya from Wilhelm Barents' time. Barents sailed along the coast of Novaya Zemlya in 1596.

In 1607-10 Hudson searches for the Northwest Passage.

In 1725-42 Vitus Bering explores the coast of Siberia and the Bering Strait. James Cook travels in the same area in 1778. Bering's expedition explores the whole of the coast with ship and dog sled and confirms that there is a Northeast Passage.

In 1818 Ross searches unsuccessfully for a navigable Northwest Passage. During this period, it is with great certainty confirmed that there is no navigable Northeast or Northwest Passage to be found. It is also the prevailing opinion that the North Pole is a continent. The last voyages were encouraged to a great extent by large rewards from the British Queen, GBP 20 000 for the Passage and GBP 10 000 for the North Pole. It has subsequently been proved that this most active period with attempts to find new seaways took place at a time when the climate was considerably colder than would have been the case both before and after. Therefore, the expeditions often did not get very far.

During the years 1819-20 Franklin and Peary explore the Northwest Passage from their different places and conclude that the Passage is to be found.

The first sailings through the passages in the North

After a relatively quiet period, the Swedish scientist A.E.Nordenskiøld is encouraged by a successful voyage to Yenisey in 1875. During the years 1878-79 he completes a successful voyage through the Northeast Passage with *Vega* (Figure 1.3). Nansen also sails parts of the Passage when he starts his drift over the Arctiv Ocean with *Fram* in 1893 (Figure 1.4). During this period the first flights over the Arctic Ocean are being planned. The first is the Swedish engineer Andrèe who takes off in a balloon from Svalbard in 1897. Andrèe and the other two who are along crash on the ice at 82°55'N after a short time. The remains are found by the norwegian *Bratvaag*-expedition on Kvitøya in 1930.

Figure 1.3
The drawing shows Nordenskiøld's Vega which was the first ship to sail through the Northeast Passage (1878 - 79). Vega was equipped with a 60 h.p. steam engine.

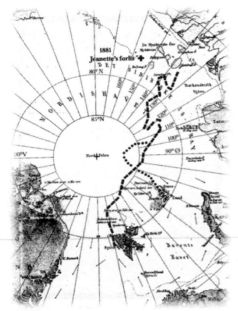

Figure 1.4
During his drift over the Arctic Ocean with Fram, Nansen reached N 86°14', and established the Trans-Polar current. The dotted line shows the route Nansen and Johansen took on skis.

After the operation over the Arctic Ocean, the Captain of *Fram*, Otto Sverdrup, took the ship on an expedition to the Canadian Arctic. During the years 1898 – 1902 he charted and annexed large tracts of land. These areas were therefore in principle Norwegian up to 1929 when Canada purchased the areas for C$ 67 000.

Despite several voyages along the Siberian coast, the large group of islands Severnaya Zemlya were not discovered until 1914. The Russians were then about to finalise a large charting expedition with the icebreakers *Taymyr* and *Vaygach*.

In the northwest, it is Roald Amundsen who is to be the first. During the years 1903-05 he was on an expedition where he also charted the magnetic north pole. On this expedition he, with *Gjøa,* was the first to sail through the Northwest Passage. That was the first of a series of expeditions that Amundsen completed. In 1911 he used *Fram* in his expedition to the South Pole, and in 1918 – 20 he sailed the *Maud* through the Northwest Passage. In 1925, Amundsen and Riis-Larsen reach 88°N by aircraft. The year after, Amundsen flew with Nobile in the airship *Norge* from Svalbard, over the North Pole to Alaska.

During the following years, expectations were driven sky-high. Charting was intensified and the areas were continually attached greater strategic significance.

The Siberian coast was completely closed to foreigners in 1917. The first formal permission for foreign voyages was not given before 1991.

The North Pole was not reached before the Americans had found methods to nagivate their nuclear submarines under the arctic ice. In 1958 *Nautilus* passed the pole point, but the year after, *Skate* was the first ship surface on the top of the world (Figure 1.5).

Figure 1.5
In 1959, the American submarine Skate was the first ship to sail on the surface of the North Pole.

The Antarctic

The Antiquity world described a country where *antipodes* lived – those who walk with their feet opposite ours. In this world it was apparently baking hot.

At the end of the 1500s, seafarers started bringing back another picture of the world.

In 1405-33 the Chinaman Cheng Ho sailed the Indian Ocean and on the coast of Africa. 64 years later, Vasco da Gama rounded the Cape of Good Hope.

In 1520, Magellan sailed the strait that is named after him. 94 years later, Schouten and Lemaire round Cape Horn.

In 1675 the Spaniard Antonio de la Roche made the first reliable observation of land in the Antarctic.

In 1819-20 Bellingshausen sailed around the Antarctic continent.

On 15 December 1911 Roald Amundsen is the first person at the South Pole after his famous race with Scott.

As early as 1904, Norway among others was about to build up the whaling activity in the Southern Ocean. Hunting lasted for many years, and at its height there were 9 factory ships and over 100 whaling vessels in activity – only from Norway. Today all whaling in the Antarctic is prohibited.

Today's limited shipping activity in the Southern Ocean is related to research and expeditionary cruises and some fishing. However, there are great expectations regarding future exploitation of the enormous occurrences of krill (Figure 1.6). Research stations have been established by several countries on shore and supplies are normally brought there by ships.

Figure 1.6
Trawling for krill in Antarctica is about to become a commercial industry. Here is the norwegian Atlantic Navigator (Photo: Knotten).

1.2 Development of the present activity in the Arctic and Antarctic

Norway: Norwegian shipping in the Arctic has developed gradually. Fishing and hunting has held a central position and during the last few years there has been more encroachment into the drift ice zone in the hunt for resources. At Svalbard, it is especially the shrimp trawlers that have penetrated the ice most often. Sealing has been reduced considerably over the last few years, amongst other things because of international pressure and difficult market conditions. Today (2011) Norway has only 2 - 3 sealers in seasonal activity, but several have applied for licences to carry out this type of activity, so it is possible that there will be a few more ships to be seen in the future.

It can be said that there has been very little research and charting activities compared to our neighbours in the Arctic. It has long been official policy not to build icebreakers ourselves, but hire them from neighbouring countries when the necessity has arisen. Safety readiness has been in place with the Coast Guard which had ice-strengthened ships (the *Nordkapp* Class), which, even though they were not specially well-suited to operate in heavier ice, were included. In 2002, however, the Coast Guard icebreaker KV *Svalbard* was commissioned (Figure 1.7). This is a medium-sized icebreaker which is further described in Chapter 5. In addition to Coast Guard duties, the ship is also used for a good number of research expeditions, e.g. for the Norwegian Polar Institute. Independent of the state research activity, Rieber Shipping in Bergen has for many years managed a fleet of ice-strengthened ships which have operated both in the Arctic and Antarctic.

Figure 1.7
KV Svalbard, Norway's only state icebreaker.

During the last few years, the oil companies have invested large resources in research in order to prepare themselves for oil and gas recovery in the Arctic. This stretches from the use of special ships with a high ice class, drilling rigs for cold climates, etc.

Experiments have also been carried out with towing seismic cables behind the icebreakers, as well as that icebreakers have been attempted used for exploration and development activity. The boom in offshore activity over the last few years has also led to that Norwegian shipbuilding has developed and built a series of ice-breaking offshore vessels. (Figure 1.8).

Figure 1.8
The Pacific Endeavour was built by Aker Brattvåg in 2006, and is a UT-758 design (Rolls-Royce).

Tourism in the form of expedition trips with ice-reinforced ships also looks to be more and more popular during the summer months. An example of this is an arrangement which is organised by the Coastal Express services, where there are ships in the Antarctic in the winter and at Svalbard and Greenland in the summer. (Figure 1.9).

Figure 1.9
The Coastal Steamer Fram is ice-reinforced, and built for expedition tourism in ice-covered waters.

The number of visits to Svalbard in 2006 was as follows:

- 51 visits of overseas cruise ships. From 2000 to 2007 the number of passengers on these ships was almost doubled from approx. 25 000 to approx. 48 000.
- 22 ships operate shorter expedition cruises (typically one week per trip).
- Two ships operate day tours (annually approximately 10 000 passengers).

The cargo ship industry is marked by the large volume of coal which is shipped from the Svea mines while there is also a considerable number of smaller general cargo ships which ensure supplies to the island group.

In connection with the renewal of the international law of the sea (UNCLOS), Norway received its economic zone expanded in 2009, along with several other states that border on to the Arctic Ocean. This was connected to the possibility of demanding the right to the continental shelf also outside the former 200 nm limit. The new possibilities have entailed larger activity within charting and geological explorations (Figure 1.10).

Greenland experienced a tragic shipping accident on 30 January 1959 with many common features of the *Titanic* accident. The passenger and cargo ship *Hans Hedtoft* (Figure 1.11) collided with drift ice near Cape Farewell and sank. All the 95 who were on board lost their lives. The ship was heavily ice-reinforced and it was constructed so that it could not sink. The year after , the *Hanne S* sank at roughly the same place and under similar circumstances. In this accident, all the 18 crew members lost their lives. The two accidents entailed that Greenland introduced several safety measures, including strict requirements regarding reporting of voyages in this area (GREENPOS).

Otherwise Greenland has had a similar development as Norway as regards cruise

traffic. For the past few years, the annual growth has been 30-40%. The tendency is also that the ships which visit Greenland are getting continually larger. As with Svalbard, there is also relatively great activity with ice-reinforced cargo vessels which supply the many settlements along the coast. Much of this traffic is undertaken by Royal Arctic Line (www.ral.dk). High raw material prices on the world market have also led to several new mining projects with requirements for shipping out – this e.g. also includes the chrysolite mines innermost in Fiskefjorden (north of Nuuk). Great attention has also been focused on possible oil and gas reserves both at East and West Greenland (Figure 1.10). Greenland has no specially built icebreaker but new Coast Guard ships are ice-reinforced to tackle the most normal ice conditions (Fig. 1.12).

Figure 1.10
Vidar Viking about to take drilling tests at Northeast Greenland (76°30'N) in 2008 (Source: Viking Supply).

Figure 1.11
Hans Hedtoft was lost on her maiden voyage to Cape Farewell in 1959 and 95 people lost their lives.

Figure 1.12
The Coast Guard ship Knud Rasmussen
carries Ice Class 1A and is 71m long.
Design is Rolls-Royce NVC-810.

<u>Sweden and Finland</u> (incl. the Baltic) has
naturally enough been forced to concentrate
efforts on ice-going vessels and icebreakers
to sail its waters in the winter. In that respect
the countries have also been leading in the
development of ice-going ships, as well as
development of regimes for safe operation
of ships in ice. In the winter readiness is
coordinated, while in the summer the vessels
are used for expeditions to the Arctic
/Antarctic, or rented out to the Navy or for
offshore purposes. In order to better utilise
the tonnage, several of the ships are operated
by Norwegian offshore shipowners in the
summer. For example, DSND operates the
three icebreakers *Fennica*, *Nordica* and
Bothnia together with the Finnish Maritime
Authorities (Figure 1.13). Similarly, an
agreement has been concluded between
Viking Supply (Figure 1.10) and the
Swedish Maritime Authorities regarding the
operation of three combined anchor
handling ships and icebreakers (Figure
1.13). Sweden's largest icebreaker, the
Oden, has in addition played an important
role in international expeditions both in the
Arctic and Antarctic (Figure 1.15).

Finland especially has concentrated great
efforts on its technological environment and
through the Aker shipyards is possibly the
largest supplier of ships carrying high ice

class in the world. The environment has its
own ice laboratory and has built a large part
of the former Soviet (now Russian) ice-
going fleet. One result of this activity is the
development of the revolutionary "Double
Acting" technology where ice-reinforced
azimuth propellers are used actively in
icebreaking and manoeuvring (Chap. 5).

Over the last few years, the traffic in Gulf of
Finland has experienced considerable
growth in tankship traffic since the Russians
opened the terminal in Primorsk. Providing
assistance to the large tankships has been a
great challenge for the relatively narrow
icebreakers (Fig. 1.14).

Figure 1.13
Norwegian shipowners operate combined
icebreakers and offshore vessels. The
offshore vessel Fennica is an icebreaker in
Finland during the winter. The ship is
116m long and is equipped with 10MW
engines with azipod propellers.

Figure 1.14
The icebreaker Tor Viking assisting a
tanker off Estonia (Photo: A. Kjøl).

Germany, Great Britain, Japan, et. al
have all concentrated efforts on building
icebreaking research ships to strengthen
their environments and make their mark in
the arctic and geopolitical context (Figure
1.15). In addition, the British had a
submarine at the North Pole for the first
time in 1971. Great Britain also has
considerable competence through the British
Antarctic Survey (BAS).

Figure 1.15
The photo shows the Swedish icebreaker
Oden (closest) and the German Polarstern
at the quay in Tromsø before they started
their successful voyage to the North Pole in
August 1991. This was the first voyage with
diesel- powered vessels to the North Pole.

The US and Canada with their large oil
and gas resources in the northern areas have
concentrated great efforts on research in the
northern areas in order to exploit them.
(Figure 1.16). Oilfield has been developed
in ice-covered areas on both the southern
and the northern coast of Alaska. Canada
has developed the fields *Hybernia* and
Terra Nova off Newfoundland where both
are located in an area where drift ice and
icebergs can occur. It has therefore been
necessary to develop platform and
production technology which can tolerate
the cold, and the ice loads. Of the more
remarkable trips which have built up
expectations, two must be especially
mentioned:

We have previously mentioned the
American submarine *Nautilus'* crossing of

the Arctic Ocean under the ice at the North
Pole in 1958.

When the US developed the oilfields off
Alaska, the question was whether the oil
should be shipped by tanker or by a pipeline
over land and down to Valdez. The oil
tanker *Manhattan* was therefore modified to
test the possibilities of shipments via the
Northwest Passage. The engine power was
increased and the bow and hull were
significantly ice-reinforced and modified.
The ship sailed the Northwest Passage early
in the winter of 1969 (Figure 1.17). The ship
managed the trip with massive help from
icebreakers, but the conclusion was that this
type of shipment was neither safe nor
financially defensible.

Both the US and Canada have developed
technology for exploitation of oil in ice-
covered regions. Canada also operates
several ice-reinforced bulk ships for
shipment of ore from the northern areas, as
well as grain from Churchill in the Hudson
Bay during the summer. The strongest of
these ships is the OBO carrier *Arctic*, which
is also equipped with very advanced ice
monitoring and navigation equipment in
order to serve the mining activity in the
Canadian Arctic outside the most ideal
summer season as well.

Figure 1.16
Oil drilling vessels in the ice off the coast
of Canada (Source: Canmar).

Figure 1.17
In 1969 the tanker Manhattan tested the possibility for shipping oil from Alaska via the Northwest Passage.

Figure 1.18
The map shows the route and daily legs of the Polar Sea and Louis St. Laurent on the voyage to the North Pole in 1994.

In 1994 the flagships in the two countries' icebreaker fleets, the *Polar Sea* and the *Louis St. Laurent* started out on a joint research mission towards the North Pole. As will be seen from the daily legs in Figure 1.18, progress was very slow northwards because of difficult ice. In addition, the American ship, the *Polar Sea,* sustained serious propeller damage.

Limitations in bunkering capacity was also about to become a problem when they encountered the Russian nuclear icebreaker, the *Yamal*, at the North Pole. Luckily, the *Yamal* could escort the two ships out into open sea and safety on the west side of Svalbard. Later (2005) the newest American Coast Guard icebreaker, the *Healy* has made approximately the same trip with the Swedish ship, the *Oden*. On this trip, mapping of the ocean bed was performed for the first time with modern multi-beam echo sounders.

In addition to its northerly activity, including supplying remote communities, specially built ships for serving research stations in the Antarctic have been built. The Americans have annual missions with icebreakers and tankers to supply the McMurdo base.

The same as Sweden and Finland, Canada especially has invested great resources for development efficient and secure rules for the operation of ships in its Arctic regions (ref. Chap. 3). Some relevant Internet links to shipowners and organisations that work with ice navigation can be:

http://www.fednav.com
http://www.icefloe.net/
http://www.uscg.mil/datasheet/
http://www.ccg-gcc.gc.ca/
http://www.tc.gc.ca/marineSafety/menu.htm

Russia (formerly The Soviet Union) is the country in the world which is most dependent on an Arctic shipping industry and is without doubt the country which has concentrated most efforts on building up the knowledge and materiel which is required for it. During the last few years, there have been a lot of efforts concentrated on the development of the oil and gas fields in Northwest Russia and in the East at Sakhalin.

An enormous northern coastal area, which stretches over 165 degrees of longitude, was closed to all international traffic after the revolution in 1917. At this point mention can be made that the Norwegian businessman Jonas Lid played a central role in the development of the timber industry in the Yenisei River. The development of the Northeast Passage, or the Northern Sea Route (Sevmorput), which the Russians call the Route, was attached such great significance that a separate Ministry was to be responsible for the development. Icebreakers, navigation, the communications system and charting were systematically built up. From 1937, establishment of research stations on drifting ice islands was started. There has almost always been at least one such station operational. Today (2011) the results of 38 such drifting stations are available. The information from these drifting stations and a range of stations on shore has contributed to increase the understanding of ice movements, climatic and hydrographic conditions in the whole of the Arctic Ocean and along the Siberian Coast. When we say that the route was closed to all international traffic, there are some exceptions. In 1940, before USSR was part of WW-II, the German Navy vessel *Komet* was given permission to sail there. The ship spent only 14 days on the trip and when it entered the Pacific Ocean, this was the start of the Nazis' war in this area. Another exception is the use of foreign timber ships in the traffic to Igarka on the Yenisei River. During the Suez crisis which started in 1967 the Northeast Passage was

for a short period offered to international users but this led to a strained political relationship with Egypt and was stopped after a short period – before any foreign ships had sailed there.

The Siberian Coast is rich in resources and the shipping industry is to a large degree built up to exploit them. Before recent years there has been relatively little through traffic, even though this was attached great strategic significance in the past. Since the distance to Japan from Northern Europe is only half of the Suez alternative, approx. 6 000 nm compared to 12 000 nm, it is naturally enough the through traffic that opens the large perspectives for the international shipping industry (Kjerstad, 90).

To give an impression of the systematic concentration of efforts, a chronological listing is given here of the events that have been of significance to the development of Russia's Northern Sea Route (the Northeast Passage) in the years following 1960.

1960: The world's first nuclear powered surface ship, the icebreaker *Lenin*, is commissioned to the escort service in the Northeast Passage (Figure 1.19).

Figure 1.19
The nuclear icebreaker Lenin in the Kara Sea in the 1960s. The ship was in use until 1989.

1970: The *Lenin* escorts cargo ships to Dudinka on the Yenisei river in winter.

1975: The *Arktika*, the world's most powerful icebreaker is commissioned. The ship is nuclear powered and equipped with 75 000 hp.

1977: The *Arktika* is the first surface ship to the North Pole, 17 August (Figure 1.20).

Figure 1.20
The Russian nuclear icebreaker Yamal
(one of six Arktika Class) on her way to the
North Pole.

1978: The nuclear icebreaker, the *Sibir* escorts a merchant ship over the Arctic Ocean, north of the island groups, in May-June.

1978: Establishment of year-round traffic on the route Murmansk - Dudinka with bulk carriers (Figure 1.21).

Figure 1.21
Russian bulk carriers at the quay in
Dudinka to load ore from the mining town
Norilsk (approx. 80km to the east).

1982: The first SA-15, ice-going merchant ship, is launched in Finland. During the years up to1987, 19 such highly ice-strengthened ships were built in Finland.

1983: Kosmos 1500 satellite is launched and contributes to better ice monitoring.

1983: 51 ships are caught in difficult ice conditions on the coast of Chukotka in October - November. Icebreakers get 50 out, one is lost.

1984: Six voyages from Japan and Vancouver to harbours in Europe (East) take place via the Northern Sea Route. Five of them by SA-15 vessels.

1985: Three SA-15 ships run experimental voyages Vancouver - Arkhangelsk in November - December.

1986: The nuclear-powered LASH ship *Sevmorput* is built to go into service from Vladivostok to eastern harbours in Siberia.

1987: The Soviet leader Mikhail Gorbachev indicates an opening of The Northern Sea Route for western ships in his legendary Murmansk speech.

1988: The first of the two nuclear-powered shallow water icebreakers, the *Taymyr*, is launched in Finland. Both are operative in the autumn 1990.

1989: The first cargo from Western Europe is transported by Soviet ship to the East. Hamburg - Osaka is sailed in 22 days with an SA-15 ship.

1990: The nuclear icebreaker *Rossia* makes a trip to the North Pole with tourists. The trip is repeated with the *Sovetskiy Soyuz* in 1991, and later with the *Yamal* (Figure 1.20) and *50 Let Probedy*. Totals 79 trips until now (2011).

1991: The French ship *L'astrolabe* sails Tromsø - Japan via The Northern Sea Route in August. Murmansk, Igarka and Providenya are visited. The Northeast Passage is thereby reopened to international traffic. (Figure. 1.22).

Figure 1.22
The French ship L'astrolabe at the quay in Tromsø 1 August 1991 just before departure to the Northeast Passage.

1991: The Soviet Union is dissolved and the development within Russia delays many projects.

1997: The Finnish tanker *Uikku* leaves Murmansk on 3 September with an oil cargo. On 15 September she passes the Bering Strait. This is a link in the co-operation with Russian shipowers and testing of azimuth propellers.

2002: The 47-foot French sailing boat *Vagabond* sails the Northeast Passage in one season. This is the start of a range of voyages undertaken by small sailing boats, but not all succeed. Vagabond sails the Northwest Passage in 2003.

2003: Shipment of oil from the Varandei terminal starts (Figure 1.23).

2006: Year-round production of oil from the fields at Sakhalin in Eastern Siberia.

2006: New ship types with "Double Acting" technology are put into service on the route between Murmansk and Dudinka. The same year the nuclear icebreaker *50 Let Provedy* ("50 Years of Victory") is

commissioned. This is an upgraded and modernised version of the Arktika Class.

2007: The Norwegian offshore vessel *Tor Viking* sails the Northeast Passage from Alaska to the Barents Sea in October. Is assisted by the Kapitan Dranitsin and a nuclear icebreaker.

2008: The nuclear icebreakers are transferred from Murmansk Shipping Company to the Russian nuclear energy authorities ROSATOM (www.rosatom.ru). Operation takes place from Atomflot in Murmansk. The two eldest ships (the *Arktika* and the *Sibir*) are phased out, and Sevmorput is evaluated modified to a drilling ship. Left behind are the icebreakers: the *Rossija,* the *Sovetskiy Soyuz,* the *Jamal* , the *50 Years of Victory,* the *Taymyr* and the *Vaygach.*

2009: The two German heavy-lift ships, the *Beluga Foresight* and the *Beluga Fraternity* sail the Northeast Passage from Vladivostok to the Ob Gulf in September. These are the first ships with bulb-bows which the Russians permit to sail through-passage (Figure 1.24). The Beluga paid € 68 000 for through-passage (must be regarded as an "introductory offer") and had worked for eighteen months to obtain the permit. Almost 500 documents were signed.

Figure 1.23
Relatively small tankers load oil in the ice-covered waters at Varandei, and then transfer it to larger ships in ice-free waters.

Figure 1.24
The "Beluga Foresight" is escorted by the
nuclear icebreaker "50 Years of Victory".

2010: The largest ship ever on the Northern
Sea Route, the 100,000dwt aframax tanker
SCF Baltica carries gas condensate from
Murmans to China in August. The panamax
bulk carrier *Nordic Barents* take 41,000 ton
iron concentrate from Kirkenes (Norway) to
China in September. The route goes north of
Novaya Zemlya and the New Siberian
Islands. For the first time ever both passages
(NE- and NW) is sailed in one season. This
is done by two small sailing vessels from
Norway and Russia.

In future years it will doubtless be the
activity surrounding the oil and gas industry
which will be the driving force for ship
activity (Figure 1.25), but the new
possibilities for transit traffic will also be
improved as a result of better ships and
warmer climate. How extensive the
international traffic will be is uncertain and
will depend on how large fees will be
demanded for icebreaker support. Today
(2011) this is approximately USD 50,000
per day.

It is first and foremost the areas in
Northwest Russia, along the Northeast
Passage and off Sakhalin where the
challenges regarding ice will be greatest, but
also in the Northern Caspian Sea there could
be significant challenges regarding ice in the
winter.

There is also reason to believe that there will
be considerable growth in various forms of
cruise traffic such as we have seen at
Svalbard and Greenland. Up until today
(2011), cruise traffic has largely been
limited to expedition tourism with Russian
ships, but gradually larger cruise ships will
find the way into this exciting area.

Many of the large icebreakers which are
now in use in Russia were built in the 70s
and 80s. In order to meet the expected
increase in the activity in the Arctic, an
extensive program was therefore started for
upgrading of the fleet. In this lie three
different concepts, where the largest class
will be nuclear-powered and have a length
of 176m and a displacement of over 32,000
tons (Peresypkin & Tsoy, 2006). Atomflot
which operates the nuclear icebreakers
asserts that the future nuclear icebreaker will
have two different operational draughts
according to whether it is to operate in rivers
or in the Arctic multi-year ice. Of the first
ships in this program was the diesel-driven
icebreaker *Moskva* which was
commissioned in 2009. The ship is 117 m
long, has two azimuth propellers and a total
engine power of approximately 16,000 kW.
The ship is classified in Ice Class LU6.

Other countries
There are several countries in addition to
those mentioned above, which can be
challenged by ice or which have research
activities in Polar Regions for periods of the
year. In the northern hemisphere this will
first and foremost be Japan, Poland and the
Baltic states. As a result of ice in both the
northern Black Sea (the Azov Sea) and the
Caspian Sea, the Ukraine and Kazakhstan
will also have to handle ice-covered
harbours. In the Southern hemisphere the
activity will largely be connected to the
research activities and readiness in the
Antarctic. Australia, South Africa, Chile and
Argentina have therefore built ships with a
high ice class for this purpose. Most of these
ships are built in Finland (Figure 1.26).

Figure 1.25
The map shows known oil and gas fields in the Barents Sea and Kara Sea.

Figure 1.26
Drawing of the ice class research vessel belonging to the South African Ministry of the Environment. The ship is 134m long and will be delivered by STX in Finland in 2012.

Antarctic
As mentioned previously, several countries have a long seafaring history in the Antarctic. Initially, this was connected to whaling and research, which in its turn led to the annexing of tracts of land. Norway annexed e.g. Queen Maud Land in 1939 after there were rumours that a secret German expedition was on its way to the area. Norway's annexation was based on charting amongst other things, which had

been done with the *Norvegia* and the *Thorshavn* expeditions in 1929-31 and 1936-37 respectively. Whale hunting has now been stopped but several countries operate extensive fisheries with trawl and long-line.

The Antarctic Treaty came into force in 1961 as an attempt to regulate legal ramifications for the activity on a continent which no-one owns, but in which many nations are interested. The Treaty covers ocean and land areas south of the 60th parallel. The Treaty lays down that the Antarctic shall only be used for peaceful activity and that nuclear activity shall not take place. All countries shall be able to have free access to carry out scientific research. Stations and equipment can be inspected by observers who are appointed by the countries that are parties to the Treaty. The Treaty can be entered into by all countries who are members of the UN, but it differentiates between consultative and non-consultative member countries. In addition to the 12 original signature states there are now (2011) 14 countries with consultative status and 16 countries with the status of non-consultative parties. The signature states were Argentina, Australia, Belgium, Chile, France, Japan,

New Zealand, Norway, the Soviet Union, Great Britain, South Africa and the US. In 1991 rules were put into place which protect plant and animal life, regulate pollution and prevent exploitation of mineral resources for the next 50 years. Further, the area is defined as a *special area* in MARPOL and work is being done to include the area where a future Polar Code for shipping will apply. The Antarctic Treaty is stated in its entirety in the Sailing Directions (Pilots) from the British Admiralty (NP-9).

In recent years there has been gradually increasing cruise ship traffic and between 1999 and 2008 the number of passengers rose from 14,000 to 46,000. Seen in context with a number of serious cases of running aground and loss, there are special reasons for concern regarding this type of traffic (ref. Chap. 6.7). Much of the traffic is co-ordinated through the International Association of Antarctic Tour Operators (IAATO). The Organisation can also co-ordinate contact with ice pilots. For further information visit: www.iaato.org.

1.3 Questions on Chapter 1

1)
Who was the first to sail through the Northeast Passage, and when did this take place?

2)
Which ship was the first to reach the North Pole?

3)
What is the Russian nuclear icebreaker *Arktika* specially known for?

4)
What was the background for the tanker *Manhattan* sailing the Northwest Passage?

5)
Which diesel powered vessel was the first at the North Pole?

6)
Which countries have signature status to the Antarctic Treaty?

7)
Where does the Antarctic Treaty apply and what regulates it?

8)
How can you obtain an ice pilot for sailing in the Antarctic?

9)
Who is responsible for the operation of the Russian nuclear icebreakers?

2 Polar geographical and environmental conditions

In this chapter, ice and environmental conditions that can be expected in the Arctic and Antarctic be described. As will be seen from the map in Figure 2.1 there are many areas with considerable shipping traffic where one can find ice in the winter. A lot of people may not expect that one can find partly difficult ice as far south as in the Black Sea at below 45° N. If in the chapter unfamiliar ice terminology occurs, please refer to the classification of ice in Chapter 4.

In recent years there has been a lot of focus on reduction of the amount of ice and the possibility for better navigational conditions in the Arctic.

The reason for the reduction is composite, but it is a fact that reduction in ice coverage has been significant. Figure 2.2 shows a relatively extreme minimum coverage from September 2007. The figure shows that both the Northwest Passage and the Northeast Passage are completely open. What is also interesting is to see the enormous area of open sea north of the Bering Strait. This area will freeze during the winter season, and will therefore be considered to be one-year ice next season. Since we know about the transpolar drift, next summer we could have open water or one-year ice almost down to Svalbard – which is of great significance for possible future shipping.

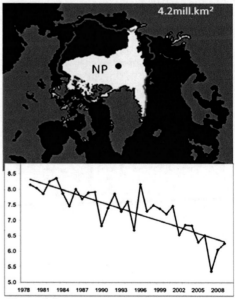

Figure 2.1
The map shows the areas where ice can be found in parts of the year and which can be challenging to shipping.

Figure 2.2
Map of the minimum ice coverage in the Arctic Ocean (September 2007). The curve shows the average in August during the years 1979-2009 in km² (Source: NSIDC).

2.1 Scandinavian Arctic and East Greenland

Ice conditions.

The ice conditions in this region are marked by the supply of warm water from the Gulf Stream. With the exception of the Balic Sea the ice limit will therefore here be pushed further north than in any other part of the Arctic. In summer one can experience open water up to approximately 82° North. However, extensive annual variations can occur as a result of how the air pressure systems are distributed. If there is relatively low pressure over the Arctic Ocean and relatively high pressure in the central Atlantic (ref, NAO-index), this will start up high west winds which push a relatively large amount of warm surface water in a north-easterly direction, and thereby contribute to push the ice far to the North. Such a pressure situation has at present (2011) been dominating since the end of the 1980s and thereby contributed to an intensified discussion regarding permanent climate changes. By studying ice spread over a lengthy period of time one has also concluded that there has been a marked reduction of the area which is covered by ice (Figure 2.3) and the thickness of the ice in the central Arctic Ocean (Figure 2.4).

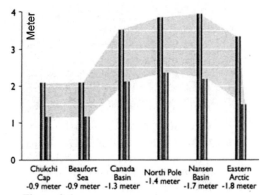

Figure 2.4
Reduction of ice thickness in various parts of the Arctic during the period from 1958-76 (columns to the left) to the period 1993-97 (columns to the right).

The reduction of the ice thickness shown at Figure 2.4 represents a reduction in the volume of the sea ice of approximately 40%. Many will then possibly ask why the sea level has not risen markedly during the same period. Upon reflection, one will see that melting of sea ice does not entail any change – consider, for example, if you have ice cubes in a glass which is then filled to the rim with water. When the ice cubes melt, the water will not run over the edge. The reason for this is that the density of the ice is lower than the water, which means that it floats, and the displaced water mass will be exactly the same as the mass of ice cubes. If it had been inland ice and glaciers which melted, the situation would have been different. Then the sea would have been supplied with "new" water and the level would have risen. What then with conditions where we hear of glacier fronts which have receded considerably – should this not lead to increase in the water level? That would have been correct if the recession represented a general reduction in the volume of the glacier. However, this is a very complex problem, and in many cases the thickness of the glaciers has in actual fact grown, even though the glacier front has receded somewhat. This is also the case in some glaciers at Greenland and Svalbard.

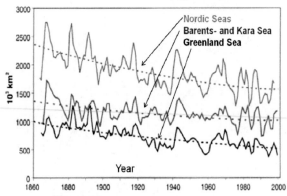

Figure 2.3
The curves shows an area (1000km²) which is covered by ice in the Barents Sea and adjacent areas over the last 150 years (Source: NPI).

Consequently, this has not led to a measurable increase in the sea level. Since we have addressed the question of measurement of small variations in the global sea level, we should mention that this is a very difficult task. This is because the metering stations could be exposed to vertical movement (isostatic uplift), as well variations in gravitation, and long periods of air pressure tendencies.

By studying the depth conditions we can see that the Fram Strait, the straits between Greenland and Svalbard, is the largest outlet for water from the deep Arctic basin. This means that there will be a transport of cold water and ice southwards along the coast of Greenland. The Fram Strait is therefore a very important area for studies of the heat balance in the Arctic Ocean and possible climate changes. Researchers from many countries therefore carry out frequent studies of the ocean currents in this area (Figures 2.5 and 2.6).

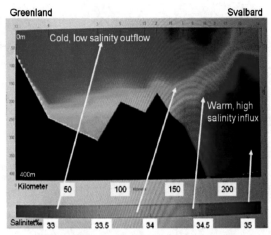

Figure 2.6
The salinity profile of the water in the Fram Strait transect (79°N). In the east the warm water with high salinity flows in to the Arctic Ocean and in the west, cold water and ice flow back along the coast of Greenland (Source: NPI).

The mixture of multi-year ice from the Arctic Ocean is therefore considerable in relation to the area between Svalbard and Novaya Zemlya. Along East Greenland it is also important to note the "ice tongue" (norw.: "Odden") which often stretches eastwards from Jan Mayen. This is caused by a turbulent current which can transport drift ice quickly and far eastwards. Transport of dense multi-year ice from the Arctic Ocean means that it is particularly difficult to sail into the coast of East Greenland. This can be a problem even for powerful icebreakers as well. The ice limits which are drawn in Figure 2.7 are mean, and large annual variations can occur (2.8). The ice in the Fram Strait will typically drift southwards at 0.5 – 1.5 knots speed (Figure 2.8).

Figure 2.5
Researcher from the Norwegian Polar Institute (NPI) measures the salinity and temperature versus depth with a CTD probe in the Fram Strait (79°N).

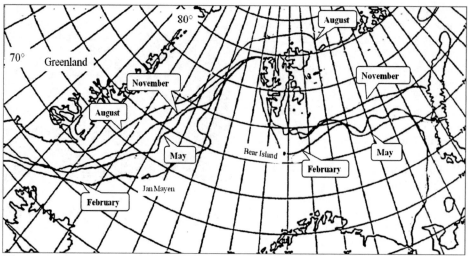

Figure 2.7
Mean ice limits for drift ice with a
concentration of 5/10 in the Barents Sea
and the Norwegian Sea.

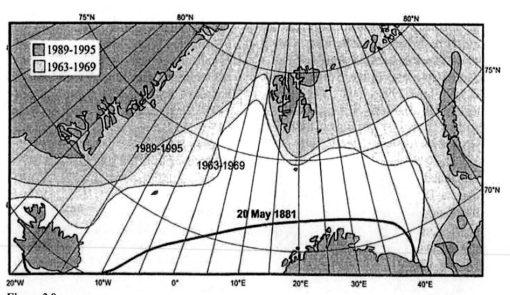

Figure 2.8
The sea ice limits at the end of April for the
periods 1963-69 and 1989-95 which
represent minimum and maximum NAO
index respectively. Extreme spreading in
1881 is also shown (Source: Blindheim
and Østerhus, 2005).

Most of the ice in the marginal ice-zone which is the transition between open sea and more permanent ice is one-year ice, but as mentioned there can also be considerable mixing of multi-year ice and "growlers" (ref. Chap. 4) which are particularly dangerous to shipping. The large massive glaciers on Greenland, and to a certain extent on Svalbard and Franz Josefs Land, (Figure 2.9) will supply dispersed icebergs. The drift of the icebergs and possible influence on marine operations has been carefully studied. Much of this research is based on satellite and aircraft surveillance (Figures 2.10 and 2.11).

Along the edge of the drift ice we also have the so-called *ice edge effect*. This is the melting of ice which causes stable water masses of between 20-30m depth. The melting will release nutrients which are caught in the ice. The combination of stable water masses, nutrients and a lot of light will during the summer months entail extensive algal blooming in a 20-50 km zone along the ice edge. Algal blooming is most dense at the edge of the ice and at the trailing edge large occurrences of animal plankton are to be found, on which fish and sea mammals feed.

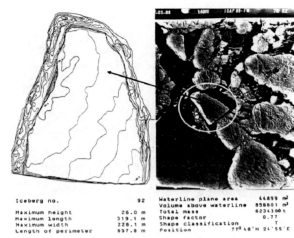

Iceberg no.		92	Waterline plane area	44899 m²
Maximum height		26.0 m	Volume above waterline	858601 m³
Maximum length		319.1 m	Total mass	6234300 t
Maximum width		228.1 m	Shape factor	0.77
Length of perimeter		857.8 m	Shape classification	T
			Position	77° 48'N 24°55'E

Figure 2.10
Aerial photo of iceberg near Svalbard taken at 6000 feet height. To the left the contours of the iceberg are plotted. The size is calculated on this basis, in this case approximately 45 000m² and 6 million tons (IDAP, 1988).

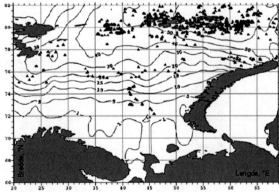

Figure 2.11
The map shows positions where tabular icebergs have been observed in the eastern Barents Sea. The lines indicate the probability in percent for ice bergs (all types).

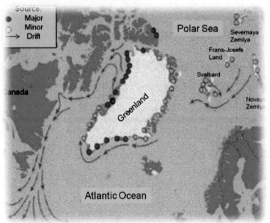

Figure 2.9
The points on the map shows glaciers which are the sources of icebergs and appurtenant typical drift patterns.

The ice warning service is at present (2011) divided into the various countries' areas (Figure 2.12) and is available on the Internet / email, fax or radiofax. In addition, in many places ice warnings will be issued in text format on navtex.

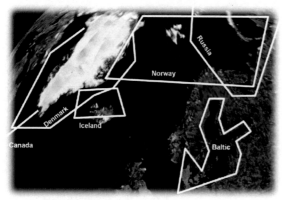

Figure 2.12
Area of responsibility for ice monitoring in the various areas in the Arctic.

There is international cooperation on ice surveillance and some countries have excellent internet distribution service with global coverage, e.g. the National Snow and Ice Data Center and the National Naval Ice Center in the US:
http://nsidc.org/index.html
http://www.natice.noaa.gov

For Norwegian areas a standard service is given by The Norwegian Meteorological Institute (DNMI), but there are also several commercial players, as for example the Nansen Environmental Remote Sensing Centre in Bergen (http://www.nersc.no) and Kongsberg Satellite Services (http://www.ksat.no) which will be able to deliver special services beyond this.

All the services are based on various forms of satellite data, everything from low resolution optical or infrared (IR) sensors to high resolution radar sensors (SAR). In addition, in some places satellite data can be supplemented with observations by ships, helicopters and aircraft (ref. also Chap. 6).

The standard service which is delivered by DNMI (www.met.no or www.yr.no) covers the area between Greenland and Novaya Zemlya. The drift ice edge and the various ice types are drawn on this chart (Figure 2.13). One will also find isotherms for the surface temperature of the sea, which can be useful for planning of voyages in arctic regions. There are also variants of the chart to be found where the patterns which indicate different ice types have been replaced by colours, but this is not suitable for distribution via fax. From the home pages of DNMI it will also be possible to get archived ice charts (http://polarview.met.no/). This can be useful for documentation after an operation or in the planning of where one wishes to look at the annual variations. It is also possible to download more detailed charts of Svalbard from the net pages. In addition, even more detailed satellite photos and chart products can be purchased. These will in many cases be relatively costly. Examples of such products are shown in Figures 2.14 and 2.15. All figures are from the area in the vicinity of the Fram Strait in May 2005.

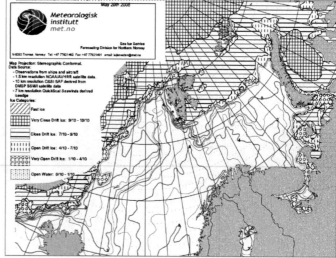

Figure 2.13
Ice chart from DNMI (2 May 2005). The chart is based on satellite data and shows ice limits, ice concentration and water temperature.

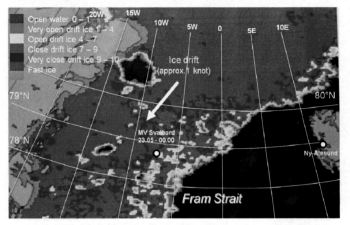

Figure 2.14
Satellite photo of the Fram Strait, May
2005. Based on data from EnviSat with
radar in "wide-mode". The dark red colour
indicates very dense drift ice.

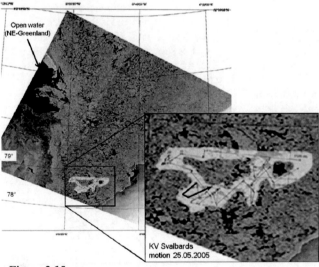

Figure 2.15
Detailed satellite photo based on Radarsat,
May 2005. The area which is lightened on
the photo is the pattern the KV Svalbard
followed for about a week by the Fram
2005 expedition.

The services which are available around
Greenland are delivered by The Danish
Meteorological Institute (www.DMI.dk).
The charts by DMI will be different from
the Norwegian ones in that they use the

international "Egg Code"
for the description of the
ice conditions in the
various areas. The
background pattern also
has the same code. For a
detailed description of this
Code, see Chapter 4. The
chart which is shown on
Figure 2.16 is typical for
this type of product, but
there are also more
detailed charts covering
smaller areas to be found.

Air pressure and wind systems

The weather situation in winter will be
dominated by a low pressure area by Jan
Mayen and a high pressure area just north of
Greenland. This will mean
that the prevailing wind
direction at Svalbard will be
east-southeast, while along
East Greenland it will be
northeasterly. In addition,
there will be a series of arctic
low pressure areas which
move quickly northeastwards
in the Barents Sea and
Danish Strait. In the winter,
the strong pressure gradient
between the polar air and the
transatlantic air will lead to
strong winds. In January,
17 % of the measurements at
West-Spitsbergen showed
over strong breeze (14 m/s).
Towards the summer, both
low and high pressure areas will be pushed
in a southeasterly direction. The wind at
East Greenland will then be dominated by a
somewhat more easterly wind. The same
will be the case at Svalbard.

8 May 2005

Figure 2.16
Ice chart of Greenland, May 2005 (Danish Meteorological Institute). Symbols are in accordance with the international "Egg Code" (ref. Chap. 4).

Climate (temperature, visibility and precipitation)
As regards temperature, Svalbard is located favourably in relation to its high latitude. The mean temperature in January is approximately -10°C. If we travel directly west to the same latitude at northeast Greenland, the temperature will be 10 degrees lower. If we follow the coast southwards along the east coast of Greenland, the mean winter temperature will gradually rise to approximately -4°C at Cape Farewell. In summer, the mean temperature will be up towards +5°C, while at Southeast Greenland there can be a mean temperature of +8°C. A typical phenomenon in the Arctic is the relatively warm air which in summer is transported over the cold sea. This will cause the typical polar fog. This

fog is low and occurs in approximately 20% of the observations surrounding Svalbard. It is often thinner at night, which makes ice navigation much easier. On an annual basis, there is little precipitation at Svalbard, approx. 400 mm. At East Greenland, the amount of precipitation will rise with decreasing latitude, and at Torgilsbu, which is adjacent to Cape Farewell the annual precipitation is almost 2000 mm.

Oceanography
Current: The picture of the current is dominated by two main currents, the warm Norwegian-Atlantic current and the cold East Siberian current which comes over the Arctic Ocean (Figure 2.17). The Norwegian-Atlantic current separates in the Barents Sea, then mixes with the polar water masses southeast and northwest of Svalbard. The cold polar current continues along the whole of East Greenland and is then called the East Greenland current. This trans-polar current which is shown during Nansen's voyage over the Arctic Ocean with the *Fram* also carries with it large amounts of Siberian timber which can be found on the coasts of Svalbard, Greenland and Iceland.

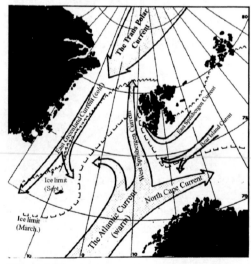

Figure 2.17
The figure shows the dominant surface current in the Barents Sea and Norwegian Sea.

Sea temperature: As mentioned above, the current picture is dominated by a cold and warm sea current. By comparing Figures 2.17 and 2.18 this condition will clearly be seen. It is especially important to note the marked division between the warm Atlantic waters and the cold Arctic waters southeast of Svalbard. This is the most important area for biomass production.

Figure 2.19
It is important to be familiar with the local conditions by studying pilot descriptions, represented here by Den Norske Los Nos. 1 and 7, issued by the Norwegian Mapping Authority, and NP-12 from BA.

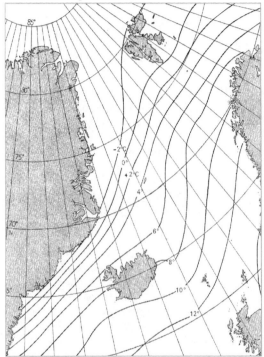

Figure 2.18
The chart shows the mean surface temperature in July (Adm. NP-11).

Tides: The tide water difference at Svalbard is normally approximately 1.5 metres. There is a somewhat less difference at Hopen and Bear Island. The times when high and low water occurs are very different and details about this are to be found in Den Norske Los, Vol. 7 (Figure 2.19). The difference decreases gradually along the coast of East Greenland, from approximately 2.5 metres at Cape Farewell to approximately 1.4 metres at Danmarkshavn which is located further north.

2.1.1 Especially about the Baltic and Gulf of Bothnia

The low salinity and lack of supply of warm water from the Atlantic Ocean makes the Baltic Sea and the adjacent ocean areas very exposed to freezing in the winter. Between Finland and Sweden the salinity will typically lie between 0 and 8‰, whilst further south in the area by Bornholm, it will typically be between 6 and 14‰. Figure 2.20 shows the ice conditions in a cold and difficult winter. From the figure we see that the use of symbols is the same as on the Norwegian charts. There are also separate charts for Gulf of Finland (Figure 2.21). On both charts we can see that the name of the icebreakers that are stationed in the various regions, are stated. The annual ice variations can be large and the degree of difficulty of navigation also shows a certain long-term tendency (Figure 2.22).

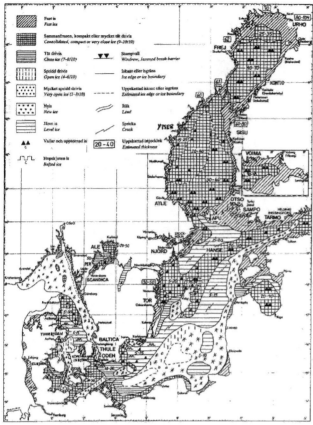

Figure 2.20
Ice chart from the Baltic Sea in a difficult
ice year (1986-87). Source: Swedish
Maritime Authorities.

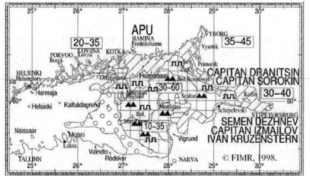

Figure 2.21
Ice chart from Gulf of Finland, December
1998, with appurtenant icebreaker
positions. Source: Finnish Maritime
Administration.

The ice in the Gulf of Bothnia will never contain multi-year ice. However, this is no guarantee that there cannot be very difficult ice conditions. It is not unusual that pressure ridges can be 15 - 20 metres thick (Figure 2.23) and together with current, the wind stress could start rapid changes and strong movement. The icebreaker services in Sweden and Finland are well equipped with icebreakers and co-ordinate all traffic in the area. In difficult conditions and a lot of traffic the icebreakers will often form convoys with several ships, possibly take a vessel in tow if there are problems with the ability to navigate. Classification of ice-going vessels, the so-called Baltic classes which DNV has notated Ice 1A, 1B, and 1C (ref. Chap. 3 and 5), also forms the basis for the icebreaker fee / pilot fee in Finnish and Swedish waters. The icebreaker administration will at all times set the requirement for the ice class based on the ice conditions. Information regarding ice conditions and current rules will be able to be obtained upon enquiry to the Maritime Authorities in Norrköping or directly to the coastal radio station in the area. On the web pages of the Maritime Authorities there are links to a range of useful information, both regarding ice conditions and rules (www.sjofartsverket.se). An extensive information service has also been established for the whole of the Baltic area on www.baltice.org. On this link will also be found training films for sailing in Finnish, Swedish and Baltic waters.

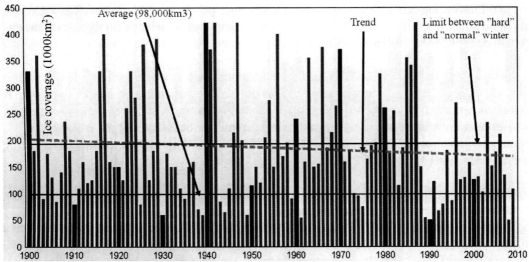

Average (98,000km3)   Trend   Limit between "hard"
                                              and "normal" winter

Ice coverage (1000km²)

Figure 2.22
The degree of difficulty of navigation in
Gulf of Bothnia here given by the ice
coverage during the years 1900-2009. The
area between the horizontal lines indicates
a normal winter. Source: The Swedish
Maritime Authorities.

Figure 2.23
The pressure ridges in Gulf of Bothnia can
be of considerable dimensions. This is from
the winter of 1990, which is regarded as an
especially favourable winter. The picture
below shows ice which has been pressed
against the foundations of a lighthouse.

2.2 Russian Arctic

The Russian north coast stretches over an enormous area, and covers approximately 165 degrees of longitude – nearly half the globe. The conditions in Northwest Russia where there are several oil and gas resources are being developed at present are largely described in the previous chapter. The further description will therefore mainly include the north coast, between Novaya Zemlya and the Bering Strait. Finally, the area around Sakhalin, on the east coast, will also be described separately.

Ice conditions

In the figures (2.24 and 2.25) the average ice limit for ice concentration of 7/10 (70 % of the surface of the ocean covered by ice). However, there are large annual variations.

In the largest part of the Northern Sea Route (the Northeast Passage), the ice will be one-year ice. However, between the northern tip of Novaya Zemlya and Proliv Vilkitskogo (Vilkitsky Strait) there can be a certain mixture of small and medium-sized icebergs. These originate from Severnaya Zemlya and the northern part of Novaya Zemlya.

When the ice has formed before the winter, often heavy packing will occur due to strong currents. If a ship is frozen in the ice there is great danger of being pushed aground in the shallow waters of this coast. The development of the ice thickness during the year is shown in Figure 2.26.

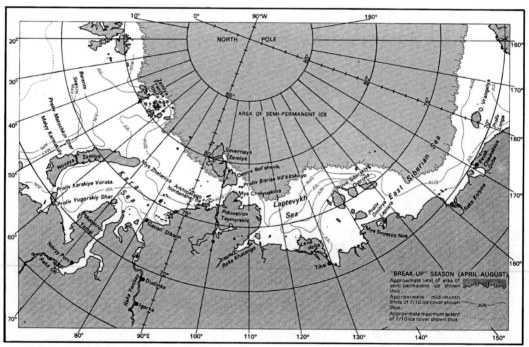

Figure 2.24
The chart shows the mean ice limit along the Northeast Passage, April – August.
(NP-10).

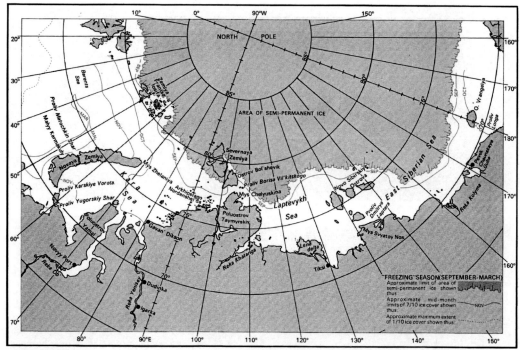

Figure 2.25
The chart shows the average ice limit along the Northeast Passage, September – March (Adm. NP-10).

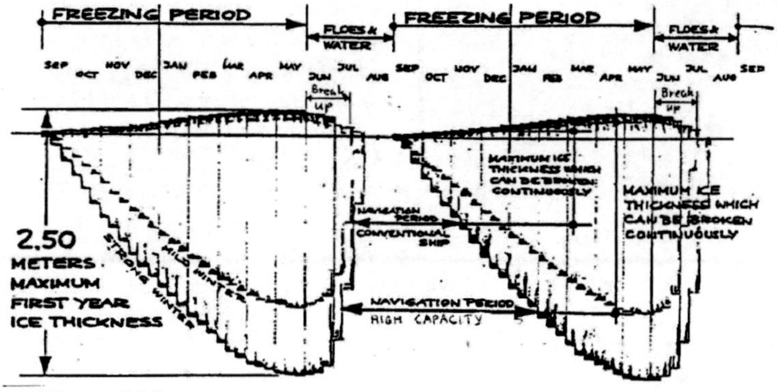

Another very difficult period for shipping is when the ice in the large rivers exits. This usually takes place in May - June. During this period the rivers are inaccessible, including for most of the icebreakers. During the winter the ice on the rivers has become approximately 1 - 2 m thick. It is covered with a snowy blanket of 1-2 m. When the water begins to rise the snow becomes flooded and re-formed to 1 metre thick white ice which covers the blue river ice. The thickness of the ice on the river is also illustrated by the sample which is taken up on the aft deck of an icebreaker (Figure 2.27).

Figure 2.26
The approximate thickness development of the ice along the Northern Sea Route during a two-year period. The figure makes it possible to estimate the season's length as a function of the ship's ice capacity (Soure: Thyssen Waas).

Figure 2.27
Sample of river ice on the deck on an
icebreaker. From the tests by IB Taymyr in
1990 (Source: Aker Arctic).

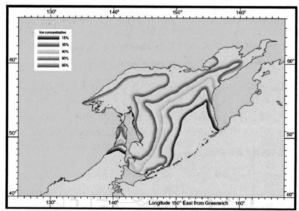

Figure 2.28
Average concentration (%) of sea ice on
the Okhotsk Sea in March (BA, NP-43).

Enormous water masses then take the ice
out at a speed of approximately 4 knots.
This course of events is largely alike for all
the north-running rivers in Siberia and
creates very difficult conditions for
shipping. Because of the extensive mixing
of fresh water in coastal waters, freezing in
the autumn could take place very quickly. It
is naturally very important to be able to
observe signs of this early on. It is especially
in the shallow Laptev Sea and the Eastern
Siberian Sea that the danger of this is

greatest. During the winter the ice
will develop a thickness of 2 - 2.5 m.
In the Eastern Siberian Sea, outgoing
ice from the rivers will often mix
with a stream of heavier arctic ice.
The result is that inaccessible pack
ice forms southwest of Wrangel
Island, just west of the Bering
Straits. In winter there can also be
very difficult conditions in the
Bering Strait and the Okhotsk Sea.
(Figure 2.28).

Air pressure and wind systems
In winter, the weather situation will
be dominated by the extensive high pressure
area over Asia. The air pressure over the
polar regions will be relatively low at this
time. In summer, there is high pressure over
the Pole and falling pressure over
Asia. In addition to this can be seen
local polar low pressure areas
which are independent of the more
global picture. In areas near Novaya
Zemlya and Severnaya Zemlya one
can also be exposed to strong
downcurrents.

Figure 2.29 shows how the
mentioned pressure picture turns
into wind. The wind roses show the
average wind dispersion by strength
and direction for January and July
respectively. It is clearly seen that
the weather situation in winter is dominated
by low pressure off the coast, which will
mean that the off-shore wind (southerly)
will be prevailing. In summer, there will be
a high pressure area off the coast and sea
breezes (northerly) will be prevailing.

It is worth noting reports from Russian
shipping companies that statistics show that
the ice situation has become gradually more
difficult for shipping in recent years (80 –
90s). The reason for this has been placed
with stronger wind systems which pack the
ice, and is probably a natural long-periodic
variation.

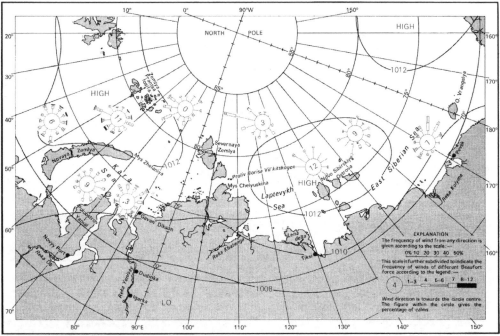

Figure 2.29
Air pressure and wind dispersion along the
Siberian coast in July. (BA, NP-10).

Climate, (precipitation, visibility and temperature).

In general, the Siberian coast is a very overcast area. The normal sailing season in July – October is the most cloudy period. Consequently, the amount of precipitation will also be the greatest in this period. The cold air, however, is not capable of taking up the very large amounts of moisture, so the total precipitation is not particularly large. The most precipitation has been experienced in the west, by Noveya Zemlya, with about 500 mm. per year. It is driest in the eastern part of the route, with about 100-200 mm. per year. The most usual form of precipitation is naturally enough snow, even though it can fall in the form of rain in the summer.

Fog or reduced visibility (visibility less than 1000 m.) can be expected on the coast for about 100 days a year. Since it is known that half of these days can be expected during

the sailing season, it is not strange that Fridtjof Nansen called the Siberian coast "the home of fog" in his books. Mist or advection fog is a phenomenon which often occurs in the vicinity of the ice edge. This creates reduced visibility, but can also help experienced ice navigators to find open channels and lanes which are suitable for sailing. This can be registered as light variations at a long distance (ref. Chap. 6). The usual frost mist which occurs when cold air blows over relatively warm water, is not usual along the Northern Sea Route. This is because of very low water temperature. The reduced visibility is most usual in the Kara Sea. At Belyy Island off the Yenisey river, 158 days of fog have been recorded in one year. Siberia is one of the coldest areas in the world. The charts of the mean temperature in winter (Figure 2.30) shows this quite clearly. During the sailing season in late summer the mean temperature climbs up to 5 – 6°C.

31

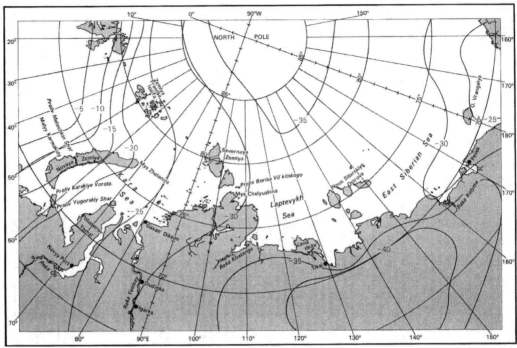

Figure 2.30
The chart shows the average temperature
along the Siberian coast in January (BA,
NP-10).

Oceanography

Current: The general current picture on the
Siberian coast is shown in Figure 2.31.
There can be seen that we have a dominance
of current which flows from west to east.
This movement is to large extent generated
by the strong North Atlantic current which
penetrates the Kara Sea with warm and
saline water. In the east we have another
current which is of great importance. This
flows northwards, through the Bering Straits
and in to a large spiral movement in the
Beaufort Sea (The Beaufort gyro).

In the summer and autumn there will be a
large outflow of water from the numerous
and large rivers. This will complicate the
current picture a good deal, and a fairly
complex composition of layers in the water
masses will occur.

Sea temperature and salinity

Table 2.1 shows the August values for sea
temperature and salinity. The chart with ice
spreading which was shown in Figures 2.24
and 2.25 will also reflect the general picture
of the surface temperature. The Norwegian
oceanographer H.U. Sverdrup, who also
accompanied Roald Amundsen on the
voyage on the *Maud* through the Northeast
Passage, did a range of extensive
oceanographic investigations of the East
Siberian continental shelf already in the
1920s.

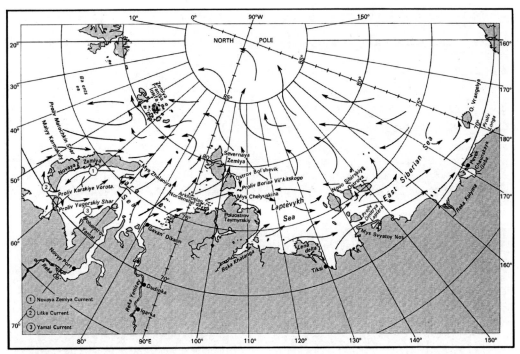

Figure 2.31
**The chart shows a generalised picture of
the surface current along the Siberian
coast (BA, NP-10).**

Table 2.1
**Typical surface values of sea temperature
and salinity in August.**

Area	Temp. coast [°C]	Salinity coast [°/oo]	Temp. ocean [°C]	Salinity ocean [°/oo]
Kara Sea	7 - 10	10	-1	32
Laptev Sea	7 - 11	2 – 3	0 – 1	30
East Siberian Sea	-	10	1 - 2	30

For cargo ships that sail with marginal
draught, this type of information will have a
certain importance for the increase of
draught and change of trim.

Tides
There is very little information available
regarding local variations along the
Northern Sea Route. It is also worth noting
that the nautical charts can operate with
differing hight datum. Datum information is
indicated on each chart. The level of the
tidewater, however, can be calculated on a
purely astronomical basis and give a good
overview. In the British tide tables and with
various software, the level of the tidewater
can be calculated for some of the selected
stations (Figure 3.32). The tables contain
information on astronomical constants along
the whole of the Northern Sea Route. It is
worth noting Z_0, which is an expression for
tidal variations around the mean water level.
The whole of the way, east of Novaya
Zemlya, this has a fairly moderate level of
0.2 - 0.6 m. On the west side, however, the
variations increase considerably, and in the
Kvitsjø area can come to values of almost 3
metres. In Murmansk, Z_0 is equal to 2.09 m.
Consequently, the total variation between
high and low water will be the double, i.e.
approximately 4 – 6m.

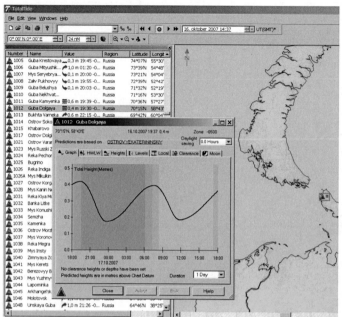

Figure 2.32
For selected stations, tides can be calculated in the electronic chart system or similar. Example from the approach to the Kara Sea (BA, Total Tide).

the Siberian coast (ref. Chap.2.2). This makes navigation very difficult for most of the year. The average ice limits are shown in Figure 2.34. It is important to note that the annual variations can be large and one must therefore always contact the local ice warning service if one is going to sail into this area. During a period in 2007, the whole of the Northwest Passage was completely free of ice. This is the first time since systematic observation was started, and many are of the opinion that it can be the start of more extensive shipping in the area. Since then we have seen the same trend and in 2010 two small sailing boats were able to circumnavigate the Polar Sea in one season – almost without encountering ice.

2.3 Alaska, Canadian Arctic and West Greenland (the Northwest Passage)

Ice conditions
The ice conditions in this part of the Arctic are often extremely difficult, both because of mixing of multi-year ice and icebergs. Most of the arctic production of icebergs takes place at Greenland and the northernmost Canadian group of islands (Figure 2.33). The supply of "warm" water is also small, which makes the melting modest even in summer. The exception in this connection is the Davis Strait, which receives some Atlantic water which presses the marginal ice limit northwards. In the Canadian group of islands the waters are shallow and the supply of water is modest in relation to what for example is the case on

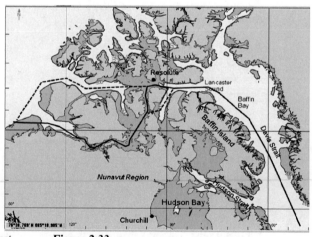

Figure 2.33
The chart shows the island group in the Canadian Arctic (Nunavut) with the most normal route through the Northwest Passage drawn in. The broken line shows alternative routes.

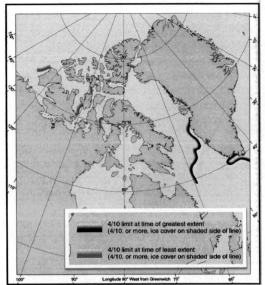

Figure 2.34
The drift ice limit (4/10) in Northern
Canada - summer and winter (BA, NP-12).

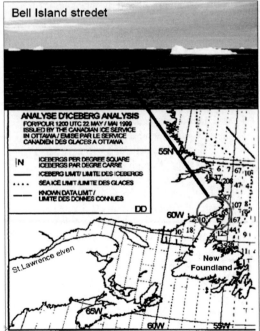

Figure 2.35
The chart shows the number of icebergs
per route (1 x 1 degrees) (May 1999). At
the same time, the picture shows the view
from the bridge of a ship.

Independent of the drift ice limit, icebergs will be able to be encountered in the whole of the northwestern Atlantic Ocean. This has always been a great danger for shipping, and was also the cause of the loss of the *Titanic* in 1912 which in its turn led to the development of the SOLAS Convention. After the loss of the *Titanic* there have been several hundred collisions, but with less loss. There is a special warning service for icebergs in the area and charts are being drawn up showing the density of icebergs (Figure. 2.35). Most of these icebergs come from the west coast of Greenland (Figure 2.9). Even though at present a considerable reduction of the glaciers is taking place, one will be able to expect almost the same iceberg density in these waters in the near future. For example, the Jakobshavn glacier on West Greenland is being reduced by approximately 30km^3 per year (2008), which in its turn will lead to a land uplift at the glacier front of approximately 15mm per year (Knudsen, 2009).

In the summer there is considerable traffic of bulk carriers which load grain in Churchill (SW in the Hudson Bay), as well as various ore cargoes in the areas around Baffin Island and the Hudson Strait. These are areas which, in periods of the year, can have very heavy ice, and where it will be important to obtain updated ice information from the Canadian ice warning service (Fig. 2.35 and 2.36). Large tidal variations also mean that one must expect large movement and variation in pressure from the ice. One shall be aware that one can often be denied breaking up the ice in areas where the Inuits use it as a transport road. For example, there will be no entry into the fjord for loading up at Deception Bay (the south side of the Hudson Strait) before 1 June for the above-mentioned reason. Even though the conditions in the St.Lawrence river and approach to the Great Lakes is considerably simpler than in the Arctic, difficult ice can also be encountered here, which means that large parts of the waterways are closed or kept open by icebreakers.

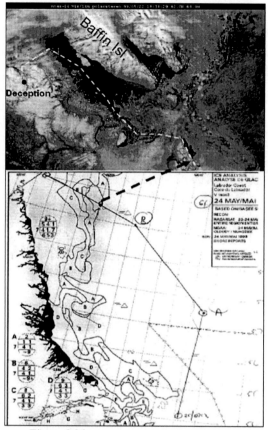

Figure 2.36
Ice chart of the approach to the Hudson Strait, with appurtenant satellite picture for May 1999.

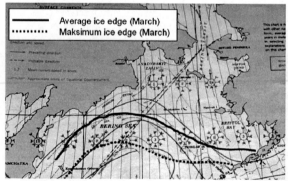

Figure 2.37
The chart shows variation in the spreading of ice in the Bering Sea (Source: Pilot Chart).

Alaska

In Alaska, in winter, the Bering Sea can be covered in ice almost completely south to the Aleutians, while in the summer there can be an ice-free area off the north coast, where oil is now being extracted (Figure 2.37). The harbours along the south coast, therefore, in winter experience a good deal of problems with drift ice. There is also an example of this at the oilfields at Cook Inlet just west of Valdez. As a curiosity, it can also be mentioned that the running aground of the *Exxon Valdez* in Prince William Sound was caused by a belt of drift ice which the navigator tried to steer clear of – an avoiding manoeuvre which created one of the greatest oil spills in history.

Air pressure and wind systems

In winter, the pressure picture is dominated by a low pressure area over the southern tip of Greenland which sets up relatively strong westerly and northwesterly winds in Northeast Canada and the Davis Strait. In the summer the low pressure area will move westwards towards the Hudson Bay, and a small high pressure area will normally be built up over Baffin Land (Figure 2.38). The pressure gradient in the eastern areas will be relatively small, and the wind conditions will therefore be calm. In the Davis Strait, the prevailing wind will turn northwesterly and southeasterly, and will normally be less than Strength 6 on the Beaufort Scale. The frequency of gales above Force 7 will not lie above 8% in any area during the summer.

Climate (temperature, visibility and precipitation)

The temperature in the summer lies above +4°C in most of the Canadian part of the Arctic, and a couple of degrees lower on the west coast of Greenland. In the Hudson Bay the summer temperature reaches 10 - 12°C. Like the other parts of the Arctic, the fog here will also be a serious problem for shipping. Dense fog usually arises when warm, southerly winds blow over the cold surface.

On the west coast of Greenland there is almost 50% fog during the summer months. Before ice is formed on the sea in winter, it is also usual that frost mist arises. The condition for this phenomenon is that the sea temperature is at least 5° higher than the air temperature. In the Canadian part of the Arctic, the annual precipitation is relatively modest, usually less than 300 mm. On the southwest coast of Greenland it can be as much as nearly 1200mm.

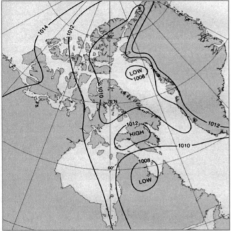

Figure 2.38
Typical pressure situation in July (BA, NP-12).

Figure 2.39
Generalised picture of the surface current in the Canadian Arctic (BA, NP-12).

Oceanography
Current: The dominant current picture is marked by the relatively warm West Greenland current which flows north along the west coast of Greenland and the cold polar current which penetrates between the Canadian islands from west to east. Both currents merge on the southeastern side of Baffin Land and form the Labrador current which transports cold water masses and ice along the east coast of Canada (Figure 2.39). The speed of the currents is relatively modest, often under 1/4 knot. In the narrowest sounds some wind directions can get up to 1-2 knots.

Tides: Along the Labrador coast and the Hudson Strait there are very large tidal differences. The largest tidal variations in the world are to be found in Ungava Bay on the southeast side of the Hudson Strait. In some places the difference between high and low water can be up to 15m. This in its turn leads to fairly strong currents and large variations in the ice pressure. When sailing in the ice, one can often experience getting stuck as a result of the ice pressure. However, this can often ease off after a few hours as the tide influences the ice.

2.4 Antarctic

Since shipping in the Antarctic is relatively modest, the description of this area will in the main be limited to the primary ice conditions. In contrast to the Arctic, ships will normally not navigate in multi-year ice such as is the case e.g. along the northern coast of Russia. Icebergs and remains of such, however, can be usual in many areas. The sea area which borders on the Antarctic lies between 60 - 70°S. These are open areas which will lie in front zones with quite a lot of low pressure in both summer and winter. The area will therefore be marked by a lot of wind and heavy seas.

Ice conditions
As shown in Figure 2.40 it is the enormous glaciers that slide into the Ross Sea and the

Weddell Sea that produce the large and characteristic "tabular" icebergs (Fig. 2.41). The glacier edge can be several hundred metres thick and the icebergs can therefore drift very far north before they melt.

Shipping in the area can in principle be divided into three categories:

- Cruise and expedition tourism. Much of this is concentrated in the areas around the Antarctic Peninsula, and the voyage takes place to a large degree in open waters and open one-year ice (Figure 2.42).
- Fishing vessels which use trawl or line normally operate in open waters, but from time to time will be surrounded by a lot of remains of icebergs (Figure 2.43).
- Research and supply ships are used to a great degree to supply the scientific stations that have been established by many countries in the Antarctic. Many of the ships are icebreakers and will be able to operate in the heaviest drift ice and sail right into the glacier fronts (the barrier) to unload (Figure 2.44).

Figure 2.40
Chart of the Antarctic. The arrows indicate the drift of icebergs out from the continent.

Figure 2.41
Tabular iceberg outside the glacier front in Antarctic.

Figure 2.42
Cruise ships without proper ice class in open drift ice.

Figure 2.43
Polish krill trawler fishing between icebergs in South Georgia (Photo: Knotten).

Figure 2.44
Research vessel the Ernest Shackleton from British Antarctic Survey (BAS) unloads materiel in the Antarctic.

2.5 Traditional sealing areas

Since the middle of the 1800s, ships from most countries in the Arctic have been engaged in sealing in different areas. Often special geographical terminology has been used in relation to the location of these grounds. The most usual grounds are shown on the chart in Figure 2.45. Today (2010) there is not much left of the commercial hunting, and this is because of more extensive protection and bad economy. Hunting has also been marked by many losses and tragedies – most caused because the ships have not tolerated the pressure of the ice. For those who wish to study hunting and descriptions of the variation of ice conditions over the years more thoroughly, Nansen (1924) and Alme (2009) are recommended. The various hunting grounds have widely differing ice conditions as described below.

current from the north. In addition, there are some icebergs which come from the north. The icebergs can often be grounded on the bottom or drift in a different direction than the normal drift ice. In this way, icebergs can constitute a great danger for a vessel which is trapped and drifts with the ice. If a ship is pressed between the drift ice and an iceberg, there is great danger that it will be lost. If one gets stuck in the ice relatively near land, there is a danger that it could be pushed into shallows where the current is the strongest. The current often flows from north to south – often at a speed of more than 1 knot. That will mean that a ship can drift up to 30 nm during a 24-hour period when it is stuck. Large differences in the tide mean that the pressure in the ice can change quite a lot within a 12-hour period. If there is wind from the northeast, there will be a danger that the ice will pack and heavy crushing can occur.

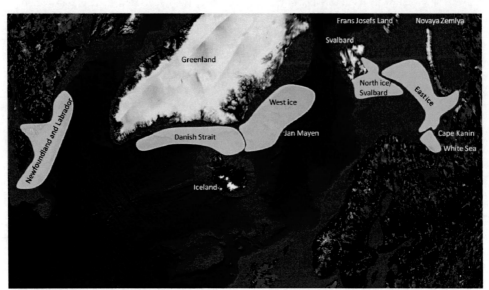

Figure 2.45
Areas where commercial sealing activity have taken place.

Newfoundland and Labrador
In winter, the waters here will consist mainly of one-year ice, but could contain some two-year ice which comes with the

Under such conditions it is important to keep close to the ice edge so that it is possible to get out. A good sign here will be to have a swell, as well as study the ice slush. If the slush is white, there is a danger of ice pressure and to get stuck in the ice.

The Danish Strait

Most of the ice in this area is multi-year ice which comes down from Fram Strait and the Arctic Ocean. In many cases there can be a mixture of one-year fjord ice with some glacial ice which comes out from the large fjords at East Greenland. It is therefore difficult and very hard ice that can impose great loads on the ships. There are also scattered occurrences of icebergs which have been calved from the glaciers further north at Greenland. The prevailing current direction goes from northeast to southwest, but in some areas with shallower banks, there can often be a more chaotic picture – which means that in some areas there will be more crushing and ridging. Normally, the current will flow at 1 knot, but it is not unusual that in some areas it can be the double. The most dangerous wind which can entail crushing is when there is a strong wind from the northeast. The wind from the west and south will on the contrary often open the ice so that accessibility is improved. Like many other ice areas, fog will often be a problem in the drift ice and out towards open sea. Normally this will improve the further into the ice one gets.

The West ice

Like the Danish Strait, the ice here consists of heavy multi-year ice which has drifted down from the Artic Sea. There can often be ice floes for stretches of several kilometres. It will be difficult to get through such large floes, even for large icebreakers. The West ice is regarded as the worst area in relation to seal hunting, and many ships have been lost. The prevailing current direction goes from northeast to southwest at approximately 1 knot, but in the case of a strong northeasterly winds and heavy crushing and ridging, there have been examples that ships have drifted at up to 3 knots. If this continues it can mean drifting of over 100nm in a couple of days. Far north in the West ice a zone is often formed between the southbound current along Greenland and the northbound current on the west side of Svalbard (ref. Figure 2.17).

This entails eddying which can pull the ice out in long tongues towards the east. We find especially such characteristic tongue on the northern side of Jan Mayen. The formation of such tongues also entail that remains of old and hard ice can travel relatively far out from the normal ice edge. This can represent a great danger for ships that sail in seemingly open waters.

The White Sea

The hunting grounds is (were) not in the White Sea itself, but sooner at the outlet – between Cape Kanin and the Kola Peninsula. The ice here is mainly relatively thin one-year ice which is pushed out of the White Sea. It is not unusual that this ice contains some harder river ice ("blue ice"). The waters in the eastern part of the strait are very shallow and can therefore entail great danger of being run aground if one is stuck in the ice. Several ships have been lost because of that. Here there is also a large tidal difference and therefore strong currents – 1 knot is usual, but cases of up to 3 knots can also occur in some places. Since there is land on several sides, strong and dangerous ridging can occur for almost all wind directions, but it will normally be most dangerous with strong winds from the northwest and north. Today there is a lot of shipping going to the harbours in the White Sea (Archangel and Kandalaksha among them) and there is a well-organised icebreaking service. The main channel for this traffic goes in the deepest water near the Kola Peninsula.

The East ice

This area stretches from the White Sea and north and eastwards. The ice is mainly one-year ice, and therefore considerably thinner than the ice found at the West ice. Some of this is also ice that has been pushed out of the White Sea. Normally, the waters here will be less exposed in relation to waves than is the case in the West Ice. On the other hand it can be very cold, which means that icing is very usual when in transit or when one is in open waters outside the ice edge.

Temperatures down to -30°C are not unusual. There is relatively little current and ridging in the area. If one is near land east of Cape Kanin, however, strong northerly winds can create ice pressure and the danger of being forced aground towards the shallow coast here. It is in this area that Russia now develops her offshore oil and gas activity and it is in the northwesterly part of the East Ice that we find the gigantic Shtokman gas field.

Figure 2.46
The Lance was built as a combined sealing and fishing vessel in 1978. Here is the ship in the Polar Institute's service outside Hornsund at Svalbard (May 2005).

The North ice - Svalbard
The waters south of Svalbard will to a large extent be one-year ice of the same type as we find in the East ice, but in some areas there can be a danger of some multi-year ice and remains of small icebergs (bergy bits, hummocks and growlers). The current will vary a good deal, and if one comes up to the Hinlopen Strait (between Spitsbergen and Nordaustlandet) one will in some cases

experience currents of 2-3knots. The prevailing current direction will be southwesterly on the east side, then turn northwards on the west side of Svalbard. This can often mean that the ice can come from the south and block the whole of the west coast of Svalbard – including Hornsund, Bellsund (Svea) and Isfjorden (Longyearbyen). The most dangerous area to travel is generally the northern side of Svalbard. Generally the current here flows eastwards and the area can for long periods be ice-free and have a broad shore channel. However, the danger is that northerly winds can very quickly put the ice down against the north coast and trap ships which are in the area. The ice will then be heavy multi-year ice from the Arctic Ocean. There are also examples of situations where a dense ice slush has occurred which has entailed that ships have not managed to get out under their own steam.

2.6 Questions on Chapter 2

1)
Name areas in the US where ice of significance to shipping is to be found.

2)
What is the smallest ice spread registered in the Arctic Sea before 2008?

3)
How much do we calculate the ice volume has been reduced by in the Arctic Sea from the 1960s to the 1990s?

4)
Why does the Fram Strait play an important role in a climate context?

5)
When can ice be expected around Jan Mayen, and in that case, what type of ice can be expected?

6)
What is a "growler"?

7)
What is meant by "the ice edge effect"?

8)
How can ice charts of the Hudson Strait be obtained?

9)
What is meant by a "pressure ridge" and how thick can these be in the Gulf of Bothnia?

10)
For how long is it probable that the Northeast Passage is open or almost open during the course of one year?

11)
Why can there be considerable brackish water along the Siberian coast?

12)
What is the typical tidal difference on the coast of Labrador and Newfoundland, and what significance does this have for ice navigation?

13)
Where do the enormous tabular icebergs occur in the Antarctic?

14)
What is it that makes the conditions for ships especially difficult in the West ice?

15)
What conditions mean that one must be especially careful in navigating in open waters on the northern side of Svalbard?

3 Description of waters and navigational conditions

The aim of this chapter is not to provide a complete description of waters for use on a voyage. It is intended as a tool in order to form an overview of the problems at the planning stage. It is recommended to always have available an authorized Pilot Description of the area in question. For further description of the problems related to electronic navigation, the textbook *Elektroniske og Akustiske Navigasjons-system* (Kjerstad, 2010*). (Electronic and Acoustic Navigational system* (Kjerstad, 2010)), is recommended.

During the past few years there has been a lot of focus on use of the Russian northern areas. Therefore, a relatively thorough description of alternative lanes along the Siberian coast is made.

3.1 Nautical charts and voyage descriptions

The datum problem.
The Arctic is a demanding area, including concerning charting. The chart quality of the largest part of the Arctic shows this. Existing charts, with a few exceptions, are in local datum (national co-ordinate system). All Russian charts are based on the Krasovsky ellipsoid (Pulkova-42/SK-42). The Russian navigational systems have also referred to this system previously, but GLONASS is based now on ITRF. It will thereby be like the global co-ordinate system WGS-84 which is used in GPS. In the remainder of the Arctic, there are other local datum (ED, NAD). However, processes have been started to convert all charts to a common co-ordinate system and global datum (WGS-84 / Euref-89 / ITRF). Then this will be adapted to the navigational systems such as GPS, Loran-C/Chayka, Glonass, and later Galileo and Compass. Transformation to the Russian co-ordinate system is relatively badly defined, and by using GPS in these areas, the absolute positioning will be relatively uncertain. Practical experiment on the Siberian coast shows that the co-ordinate displacement between WGS-84 and Pulkova-42 lies above 100 meters, both in an easterly and northerly direction.

Norway
In the Svalbard region some of the charts are based on relatively old and imprecise surveys. With the exception of the harbour charts the scale is 1:100 000 or less. The Norwegian Mapping Authority, which today is responsible for surveys, has in recent years performed surveys in the most needed areas (Fig.3.1). Progress in survey work is limited by resources and a short survey season, and it will therefore take many years before we have a good charting basis in the whole area (both electronic and paper charts). When the waters on the northern and eastern sides are very shallow, great care and vigilance is required of the navigator (Figures 3.2 and 3.3). The Mapping Authority experienced this itself when they ran their ships *Lance* aground on an uncharted shallow near Kvitøya. The responsibility for charting on the shore side is given to the Norwegian Polar Institute in Tromsø.

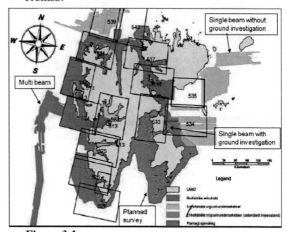

Figure 3.1
Chart coverage and survey status at Svalbard. White areas is only sporadic data, the reliability of which is limited (Norwegian Mapping Authority, 2006).

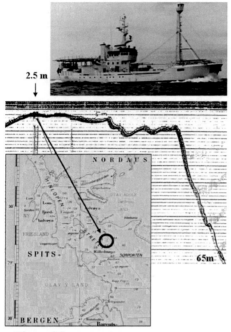

Figure 3.2
Echogram from the research vessel Lance
shows 2.5 m. In the chart of the Hinlopen
Strait, there was no indication of
dangerous shallows (Jansen, 1992).

Figure 3.3
A surveyor from the Norwegian Mapping
Authority standing on a shallow where the
chart showed 300 m.

During new surveying, it has often been found that the land masses have moved 1500 – 2000 meters in relation to the old graticule. In actual fact, it has been experienced that islands have been up to 12 km "out of position". The problems with different datum can also be experienced during navigation with modern electronic chart systems in well surveyed areas. Figure 3.4 shows an example that the chart basis of an electronic chart system (ECDIS) has different datum in different scales. The position will therefore be able to jump considerably upon change of scale.

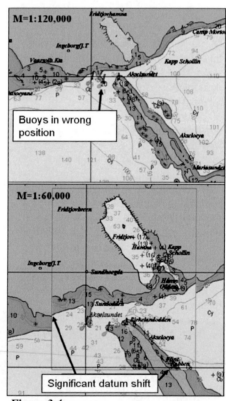

Figure 3.4
Datum change can clearly be seen on
different chart bases here in Aksel Sound
at the approach to the Svea mine at
Svalbard.

One problem which has become particularly relevant after the glaciers receded is navigation at the glacier fronts. These are areas where it is not very probable that reliable chart data is to be found (Figure 3.5) and we have already experienced that many ships have run aground in such areas. If one is to go relatively safely in to such areas, it is necessary that one either navigates with forward seeking sonar, or has a tender equipped with echo sounder which can survey the waters at a good margin in front of the ship. In 2007 there was also a very serious accident with a small cruise ship which navigated close to a glacier front. While the ship was just at the glacier, large pieces of ice fell over the forecastle head and injured many of the passengers.

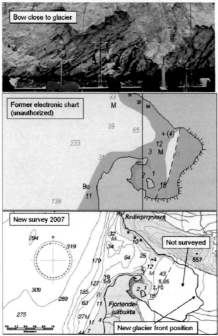

Figure 3.5
Navigation near glacier fronts. The bottom is normally not charted and the glacier can calve and damage the ship. The bottom chart illustrates the quality improvement after recent surveying (Norwegian Mapping Authority, 2007).

In 1988, the Norwegian Mapping Authority issued a description of waters of Svalbard and Jan Mayen (Den Norske Los, Vol. 7) which provides useful information. This publication is in English and Norwegian and should be kept on board all ships that sail in the area (new edition 2011?). The British Admiralty (BA) also has a pilot description which includes West Greenland, Iceland and Svalbard (NP-11). American editions can be found on the Internet (http://www.nga.mil).

Russia
Extensive resources have been spent over a long period of time to chart the Siberian Coast. The Russian charting basis which is to be found is therefore extensive. The scale is normally 1:100 000 with harbor charts down to 1:15 000. The positioning which is used in surveying work is based on microwave systems, 2 kHz systems, circular modulated Chayka (ref. Chap. 3) and satellite systems, which are largely the same as have been used in Norwegian arctic and the rest of the world. The data collection is automatic and depth measurements are based on single (after WW-II) or multi-beam (after 90-ties) echo sounders. The enormous areas and the ice have often made it necessary to have a relatively large survey line distance, which will entail few soundings and a risk of undiscovered shallows (Figure 3.6).

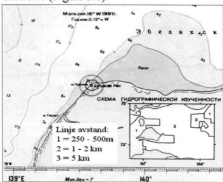

Figure 3.6
Chart of the area by the New Siberian islands. On the chart the line distance between soundings is stated. The distance can be up to 5km.

Taking into consideration that the largest part of the navigable area is shallower than 30 meters the waters will require careful navigation. For that reason, Russian navigators slavishly follow the recommended routes which are drawn up on the chart when the ice conditions allow it. The Russian charts are now available both electronically and in paper form.

American and British charts of the area have large limitations and are not suitable for navigation. One can actually find charts where islands that do not exist are to be found marked (Figure 3.7). The reason for such conditions is that low sandbanks can be washed away by the ice, and possibly new banks can be formed by the same process. There are naturally enough shallow areas and river mouths where this problem is most prominent. In shallow areas there is also the possibility that remains of pressure ridges are frozen fast at the bottom and in that way create danger for shipping. Even though the data basis is old, the description of the waters can be of good use (US-183, NP-10 and NP-23).

Canada and West Greenland.
There are usable Admiralty charts along the west coast of Greenland as far north as 76°45'N. In addition, the Danish Kort & Matrikkelstyrelsen (National Survey and Cadastre) (www.kms.dk), works continuously with improvement of the basis for charts in the busiest areas. The quality of older charts, here and in the Canadian islands, is characterized to a large extent by random observations by ships which because of the ice conditions have sailed in new areas. The positions of the soundings are therefore relatively unreliable. On the most important sailing routes and in the vicinity of harbours, systematic surveys of good quality have been performed. The contour of the land on the nautical charts is largely based on recent aerial surveys and is also of good quality (Figure 3.8). The limitations of each chart will normally be stated as a note on the chart or on the chart border. Since it is not a matter of course to have a good chart basis in the Arctic, it is important to read these notes carefully. The description of the waters in the area is covered by NP-12. There are also good national charts and descriptions of waters.

In connection with the new Law of the Sea Convention, coastal states can extend their economic zones. This requires that the states can document that the continental shelf stretches further out than 200nm. In this connection, there has recently been several expeditions where charting with multi-beam echo sounders has played an important part. Experience from these surveys indicate much of the same features we find on other continental shelves – e.g. "pock marks" which can often indicate that there are gas deposits in the area. A milder climate and steadily increasing traffic including cruise ships, will also mean that charting of more remote regions will have to be intensified in the years to come.

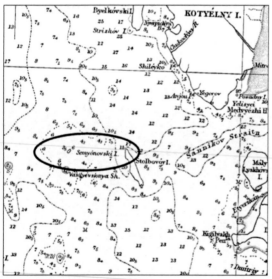

Figure 3.7
On an overview chart from BA islands are marked that do not exist.

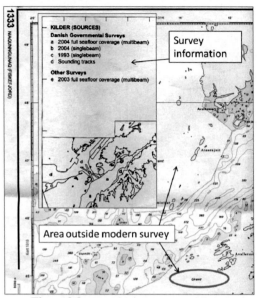

Figure 3.8
Chart example from Greenland where unreliable soundings have been removed.

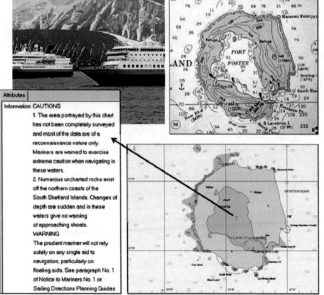

Figure 3.9
The "Nordkapp" receives assistance from the "Nordnorge" after grounding near Deception Island in 2007. In the text box, information from the electronic chart indicates that the chart is insufficient and that there are unknown shoals.

The Antarctic

Since the waters around the Antarctic are placed under national states, naturally enough not much detailed and systematic charting has been undertaken. The quality at best is at the insufficient level which is described in the most remote regions of the Arctic. However, there are charts to be found, as well as Pilot Descriptions (NP-9) from the British Admiralty. Also the US and the countries which border the Antarctic have some charts. Many of the charts are on a small scale, but at the Antarctic Peninsula there are several charts in the 1:200 000 scale which are considerably worse than those we normally use in more busy shipping areas. The charts also contain warnings regarding the substandard quality and that navigation must be undertaken with the greatest care. Even though the quality is bad, there are also electronic versions of the charts – both in raster and vector format. Such chart formats can often give navigators a false sense of security and it should absolutely be emphasized that source data is checked and the quality of this evaluated. Taking into consideration that voyages are made by larger ships, also in the Antarctic, one cannot lean on experiences that "things have gone well before".

An example of the dangerous waters was experienced by the coastal express ship *Nordkapp* when it ran aground near Deception Island on the Antarctic Peninsula in January 2007 (Figure 3.9). Running aground luckily did no significant damage/injury to people or the environment. However, one shall be aware that the margins between a tragedy and an "unsuspecting" grounding can be hairline in such regions. That the *Nordkapp* operated together with another coastal express ship, the *Nordnorge,* contributed to the outcome being relatively good. Both the *Nordnorge* and the British ship, the HMS *Endurance* came and brought passengers to safety back in Argentina. The costs of loss of income, compensation and repairs were formidable. The incident also indicates the importance of operating together with other ships.

If one is to safeguard oneself properly in such marginal waters, one must either install a forward looking sonar system, or one must sail in front of the ship with a smaller survey ship or tender with echo sounder. The chart system *Olex* contains an example of being able to transfer data from the echo sounder and GPS on the tender so that the ship can chart the waters ahead in true time.

3.2 Navigation and buoy system

Buoy system.
Because of the ice conditions, floating buoys are attempted used as little as possible. Instead, the leads are most often marked with leading lights and shore-based beacons, cairns and lights (Figure 3.10).

Floating marks in the Arctic are based on the international IALA system (Region A). Radar reflectors are mounted on a large part of the markers. Because of the difficult ice conditions in the area, one must be especially aware of the danger of movement or damage to the buoys. In winter it is usual to take the buoys ashore and replace them with markers that stand on the fast ice. For further information regarding lights and sound signals, reference is made to the "Admiralty List of Lights", Volume M, which describes what is known in this area in detail.

In areas where there are traffic separation systems (TSS), for example at the approach to Helsinki, there will be special rules when the waters are covered in ice. This is because the channels that are used by icebreakers can be located outside where TSS is marked on the chart.

In the Canadian Arctic, a modified lateral system is used (ref. NP-12), while at

Svalbard floating markers are no longer used, with the exception of the approach to Svea in the summer.

Figure 3.10
Along the Siberian coast, large shore-based markers have been built as aids to navigation. Here at Belyy Island in the Kara Sea. Racon and light are operated by a radioactive source which is located at the side of the marker.

Pilot services
The obligation to have a pilot is independent of whether there is ice on the sea or not. This means that in many cases conventional piloting and pilot boarding can be difficult (Figure. 3.11) (ref. Part II). In some cases dispensation can be granted or captains can be given special permission (Pilot Exemption Certificate (PEC)). In Norway, new forms of licenses are under evaluation for sailing into Svalbard. In that respect, special simulator-based courses can be introduced to qualify as a "Polar Pilot" when the Harbour and Waters Act is introduced in whole or in part. In areas with organized icebreaker services, this will in principle act as a pilot service. In Canada and Russia it is usual that ships take on board an ice pilot who will also guide the navigators with routing through the ice.

Figure 3.11
The pilot can arrive with a snow vehicle, as here in Finland.

Navigation systems.
GNSS: The satellite systems which are now operative (GPS and GLONASS) cover the whole of the Arctic. The performance of GLONASS, because of it somewhat higher orbit inclination, will give a little (marginally) better output at higher latitudes. GPS has for many years showed good performance in expeditions all the way to the North Pole (Figure 3.12).

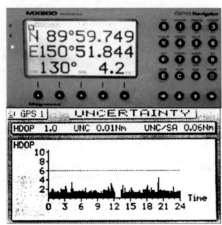

Figure 3.12
Photo of a GPS display at the North Pole. Below is shown the diagram of a typical 24h HDOP on GPS, observed during sailing through the Northeast Passagse.

 The GPS has different systems for calculation of course / heading. These

systems are based on phase measurements of the satellites' carrier wave between three or four antennas and are usually called a GPS compass (Figure 3.13). At present (2011) this system is not approved as the main compass, but is in actual fact superior to the gyrocompass at high latitudes. However, it is particularly important that the antennas are placed as high and free as possible in order to avoid multidirectional interference (multipath).

Figure 3.13
GPS based compass antennas are tested here on the Swedish icebreaker Oden while on a mission to the North Pole in 1991.

Differential GPS has partial coverage in the Arctic through the satellite-based WAAS, EGNOS and MSAS systems. In theory, these are able to transmit corrections in the whole of the area of coverage to the geostationary satellites (see also Communication). The problem is, however, that not all corrections can be transmitted to all the satellites one will see at high latitudes. Therefore, they will be rejected and the geometry in the position fix will be worse – in practice, worse than GPS in normal mode (Figure 3.14). Ground stations (RIMS) have been developed for EGNOS at Jan Mayen and Svalbard in order to expand the area of coverage somewhat. In the Barents Sea corrections will also be able to be transmitted via the Eurofix system at Bø in Vesterålen. This requires that one has a receiver that can receive the signal which is modulated on the Loran-C signal (Figure 3.15).

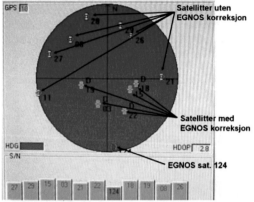

Figure 3.14
Skyplot by GPS satellites and EGNOS in the Barents Sea. Note that the satellites in the north do not receive correction signals from EGNOS.

Loran C and Chayka are two radionavigation systems which are similar, in actual fact, completely compatible. Chayka is the Russian variant of this shore-based hyperbolic system. On the western part of the Northern Sea Route, a Russian chain can be used (Figure 3.16), while in the eastern part, signals will be able to be received from both the American and Russian stations which are now co-ordinated in common chains. In the areas SW of Severnaya Zemlya and eastwards to 130° East, there will not be satisfactory coverage. Concerning datum, the same applies here as for the satellite systems. On most of Greenland and Northern Canada, Loran-C is available. In the Norwegian Arctic at present (2011) there is no satisfactory coverage, and it is uncertain how long Loran-C and Chayka will be operative – especially after Glonass and Galileo are declared operative (probably 2011 – 2013).

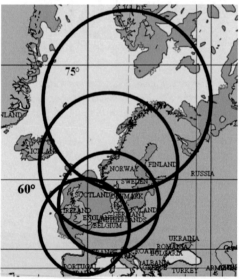

Figure 3.15
Approximate area of coverage for Eurofix. The northerly ring represents coverage by the station at Bø.

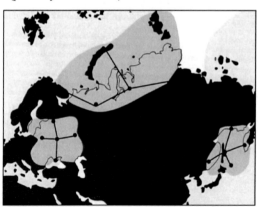

Figure 3.16
Coverage for three Chayka chains (Russian Loran-C) in the northern areas of Russia. The lines are base lines between the master and slave station.

From approximately 2013 one will also be able to use signals from the European satellite system Galileo and the Chinese Compass system. The performance is expected to be a little better than the present GPS and Glonass performance, mainly because of a few more satellites.

Radiobeacons: A range of radio beacons for navigation have been built in the Arctic (ref. Adm. List of Radio Signals). The significance of these aids has been considerably reduced in recent years. Several nations are about to reduce the number of stations, or use the stations for transmission of differential GPS signals.

Radar beacons (RACON) are an important navigational aid in radar navigation, including in the Arctic. Shipping in the Northeast Passage has good use of a range of radar beacons. It is also Scandinavian navigation policy to build up this offer gradually as the radio beacons are taken out of service.

The Gyrocompass reigns almost supreme at present for north reference in practical navigation. If we disregard the optical "gyrocompasses" all are built with one or two stabilized gyroscopes. Common to them all is that they meet the hard reality of physics when they get up to high latitudes. The problem can be divided into two:

Adjustment / stability and **speed errors on the gyro.** Stabilizing is given by the earth's rotation momentum amongst other things, which in its turn is dependent on latitude. The settling time (T) will be affected by this both at start and course changes. The time is given by the following formula (1). The speed error (E) which is characterized by the relationship between the ship's speed vector and the earth's peripheral speed will therefore also be dependent on latitude (formula 2)

1)

$$T = 2\pi \sqrt{\frac{H}{B \bullet w \bullet \cos \beta}}$$

2)

$$E = \frac{v \bullet \cos \phi \bullet 57.3}{901 \bullet \cos \beta + v \bullet \sin \phi}$$

H = The gyro's angular momentum
B = constant given by the precession weight mass
w = the earth's angular speed
v = the speed of the ship
φ = the course of the ship
E = speed error
β = geographical latitude
T = settling time

From the formulas we can see that both these unfortunate circumstances will increase heavily with increasing latitude. The manufacturer's ready-calculated correction tables for these circumstances will seldom go to higher latitude than 75°, which means that one must undertake a thorough analysis for each individual compass type. It is not unusual that in the waters north of Svalbard (81 - 82 degrees North) there can be erroneous indications of 20 - 25°. Frequent course changes which are often necessary when sailing in drift ice will lead to extensive adjustment problems for the gyrocompass. The problem is quite clear based on formula 1 which indicates that at 84° North will have an stabilizing force which is only 1/10 of the force at the equator. Experiences from voyages with the Swedish icebreaker *Oden* to the North Pole in 1991 showed that the ship's 2 gyrocompasses were completely unusable at 84° and 87° North. Some compasses can also have indicated accuracy given by a factor of 0.5 – 1, divided by the cosine of the latitude. It is then easy to see that the error will go towards the infinite when approaching 90°.

In 1996 the *Hanseatic* ran aground in the Canadian Arctic. In hindsight it became clear that the ship had made navigational errors as a result of a gyrobased bearing on the radar. Such bearing will have at least the same error as the gyro.

Inertial navigation (IN) is based on gyrotechnology and accelerometer. In theory this should be able to be used all the way to the North Pole, and it was this technology which enabled the American nuclear submarine USS *Nautilus* to cross the pole point in 1958. Advanced IN systems have today made navigation itself under the ice a pure routine for Russian, American and British submarines. In practice, these systems will also have greater uncertainty in extreme latitudes. On *Oden*'s first mission in 1991 the IN system had problems very near the North Pole.

Celestial navigation

Celestial navigation can of course be used in the Arctic, as other places on the earth. However, there can be special problems connected with the performance of the observations since there is often no clearly definable horizon because of ice. In wintertime, the dark will make ordinary observations possible only for a short period in the middle of the day. Both the above problems make it necessary to use an artificial horizon. The effect can be achieved by using a "bubble sextant" or horizontal mirror which must be adjusted on a firm ice base (Figure 3.17). In both methods, the observed height shall be corrected only for refraction, since horizon dip does not have to be included. Consequently, the total correction in found in some tables cannot be used. If one measures the angle towards a horizontal mirror, the observed angle must first be halved before it is corrected for refraction. With ordinary observation of stars, planets and the moon, an observation will be made first within a relatively short twilight period. If an artificial horizon is used this will not be of critical significance.

aid of an artificial horizon (horizontal mirror) the following star heights are read: Eltanin at 113°11.1' and Alpheratz at 71°55.1'. The sextant's index error = +0.3' and eye height is 15 meters. The temperature is -20°C and the air pressure is 1015 HPa.

Solution of example with plotting sketch:

Object	Eltanin	Alpheratz
√GHA (18-00)	299°55.3'	
(24-32)	6°09.0'	
DR long.E	14°56.0'	
	321°00.3'	321°00.3'
* SHA	90°54.3'	358°00.9'
* LHA (-360)	51°546'	319°01.2'
* Decl.	N 51°29.6'	N 29°02.6'
Sextant ht.	113°11.1'	71°55.1'
i.e.	+ 0.3'	+ 0.3'
Meas.ht.	113°11.4'	71°55.4'
div.2	56°35.7'	35°57.7'
refraction	- 0.6'	- 1.3'
temp. corr.	- 0.1'	- 0.2'
Obs. ht.	56°35.0'	35°56.2'
Calc. ht. q.	56°31.8'	35°49.9' [1]
Height error	tow. 3.2'	tow. 6.3'
True bearing:	243°	135° [2]

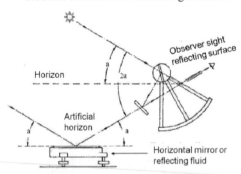

Figure 3.17

Principles of celestial observation with an artificial horizon. Note the halving of the measured angle, and that one does not need to correct for horizon dipping.

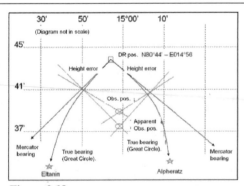

Figure 3.18

The plotting sketch shows in principle a graphic solution of the celestial observation in the example above.

Example:

You are frozen fast in the ice north of Svalbard and according to dead reckoning your position is N 80°44' - E 14°56', UTC is 18-24-32 on 21 October 1990. With the

Formulas referred to in the above calculation:

1)
$\sin h = (\sin la * \sin d) + (\cos la * \cos d * \cos LHA)$
2)
$\cos b = (\sin d - \sin la * \sin h) / (\cos la * \cos h)$

h = obs. height
la = dead reckoning latitude
d = declination
b = true bearing
LHA = local hour angle of the celestial body

Comments on the solution:
As shown by the calculation on the previous page we have divided the measured angle before we correct for refraction. If the **bubble sextant** had been used, this would not have been necessary. The procedures are otherwise alike. It is important to note that we have not corrected for dip, the observer's eye height will therefore not be of interest. Since the ordinary tables are drawn up for "normal" refraction, in the Arctic regions it is important to use additional correction because of the extreme cold. In our example, this correction was relatively small (-0.2'), but with lower heights it will be very marked, approximately 10' at a height of 10°. The uncertainty of refraction, however, means that one should avoid low observations in the Arctic.

The spherical calculation of the celestial body's bearing is often set out in the Mercator chart as shown in Figure 3.18. Since the bearing in reality is a great circle bearing, at high latitudes we will make some considerable mistakes by setting the bearing directly out in a Mercator chart.

3.3 Communication conditions and radio services

Both in the Russian, Scandinavian and Canadian Arctic, there are several available radio stations for VHF and medium wave telephony. The transmission plan for them is found in the Admiralty List of Radio Signals. Along Siberia there are also several less powerful stations, where the listening and transmission plan is published in the harbour description. In Northwest Russia, the navigational warnings are transmitted both in Russian and English from Amderma and Dikson. Amderma covers the Kara Sea and Franz Josef Land. Dikson covers the Kara Sea, Franz Josef Land, Laptev Sea and the Eastern Siberian Sea. The Navtex service is still not developed here. Ice and weather forecasts are transmitted by fax from Amderma, Mys Chelyuskina, Tiksi and Pevek. These services will normally only be in Russian.

Short wave communication is another possibility, but since both Rogaland (Norway) and Lyngy (Denmark) radio stations have ceased their services, there is now little relevance for ships from Scandinavia to use shortwave communications. The short wave is necessary for ships that shall have the GMDSS requirement met for voyages outside Inmarsat coverage (A4).

Satellite communication, via Inmarsat, will also be possible in parts of the Arctic. Because of the low elevation angle in relation to the satellite, the system will have large areas that are not covered. The coverage chart in Figure 3.19 is based on a satellite elevation angle of minimum 5 degrees, which is the guarantee for telephone communications from Inmarsat Standard B. Field tests from Svalbard and the easterly Barents Sea, however, indicate that it is possible to achieve connections with lower elevation. By transfer of data on Standard C one can actually often achieve communication when the satellite is just under the horizon. This was also confirmed by the Nansen Centre for Environment and Telemetry, which used fax on Inmarsat (Standard A) to transmit satellite photos from ERS-1 on a mission through the Northern Sea Route. If we look at the coverage chart (Figure 3.19) with 0 degrees elevation, we will see that there only a small area east of Taymyr, west in the Northwest

Passage and north of Svalbard which is not covered. In Gulf of Bothnia problems have been reported with satellite communication when there have been extremely low temperatures. If we assume that most antennas are stabilised with the aid of a gyro system, one must also expect problems with adjustment as described for the gyro compass when this is used at high latitudes.

small (up to 2.4 kbps). However, a special service is to be found, called "Open Port" which uses several telephone lines at the same time. With the aid of this service and adapted equipment, one is able to achieve a transmission speed of up to 128kbps. Detailed information of the services are to be found on http://www.iridium.com/. There are different types of equipment – both handset (Figure 3.20) and rigidly mounted systems with fixed antennas.

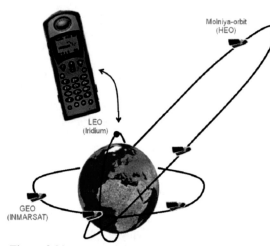

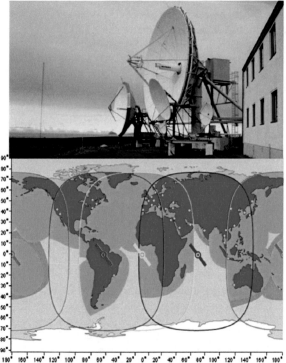

Figure 3.19
Coverage chart for the geostationary Inmarsat satellites, based on 5° elevation. The photo shows antennas at Isfjord Radio (N78°03'), and we can see that they are almost angled towards the horizon.

Iridium is a satellite communications system where the satellites go in low polar orbits (LEO). This is not a part of the ships' mandatory GMDSS equipment, but will in practice be extremely suitable for communication in all parts of the Arctic and Antarctic. It is also possible to transmit data over Iridium, but the bandwidth is relatively

Figure 3.20
Iridium telephones have good coverage in all parts of the Arctic and Antarctic. Here is shown a lightweight handset together with an illustration of various satellite orbits.

Previously, the Arctic has been excluded from the international definition of NavArea and MetArea where different states are given the responsibility of transmission of navigation and weather forecasts. Recently, the polar areas of responsibility have been approved by IMO. There were then five new areas (19 – 24) with divisions as shown in Figure 3.21.

Figure 3.21
Division of the new NavArea for Arctic areas.

3.4 Arctic Harbours

In this chapter, information will be given in brief about the harbours which can be relevant to visit on voyages to the Arctic. Harbours which are located far up rivers will not be discussed. In addition to the harbours it is worth noting that the large icebreakers will be able to assist in some areas with services connected to bunkering, piloting, towing, etc.

Svalbard
Longyearbyen. There are two quays with depths of 9 and 16 meters respectively. Both have a length of approximately 50 meters (Figure 3.22). There are possibilities for bunkering of oil and water. Information at : http://www.portlongyear.no/. There is an airport with departures to Norway and Russia. The Governor has helicopters stationed. (http://www.sysselmannen.no).

Figure 3.22
Cruise ship at the quay in Longyearbyen.

Pyramiden. Here is found a 75-metre-long quay with depths of between 8 and 9 meters. The harbour is operated by a Russian coal company and visits must be agreed in advance. There is no permanent settlement any longer.

Svea (Figure 3.23) Store Norske Spitsbergen Kullkompani (http://www.snsk.no/) operates coal mining and coal is transported with ships of up to 75,000 tons. These are ships which when loaded have a draught of approx. 14 meters. Production in 2010 was approximately 2 million tons. The approach is via the narrow Aksel Strait (Figure 3.4) and larger ships will be assisted by tugs.

Figure 3.23
Ship assisted from the quay in Svea.

Barentsburg. This mining community is operated by Russia, but the 76-metre-long quay can be visited by agreement. The depths are between 8 and 9 meters.

Ny Ålesund. This is the world's northernmost harbour (Figure 3.24). The place was previously a mining community, but today it is the base for an international research station. There are two quays – the largest of which can be put into by fishing vessels and research ships without any problem. Larger cruise ships often set passengers ashore with their own tenders.

Hornsund. This is the southernmost settlement on Spitsbergen. Here is a Polish research station which is operated on a year-round basis. There is no ordinary quay installation here and supplies must be set ashore by amphibious vehicles.

Figure 3.24
The cruise icebreaker Polarstar and the
expedition ship Stockholm in Ny-Ålesund.

Figure 3.25
The main quay in Pevek often handles 2 –
3 merchant ships at the same time, but
ships must often lie at anchor for a long
time in the roadstead to get into the quay.

Russia

Dikson. This is the headquarters for arctic
operations in the western part of the
Northern Sea Route and approach to the
Yenisei River (Figure 3.27). Shipping
activities are mainly operated by the
Murmansk Shipping Company. The place is
also a centre for the meteorological services
in the area. The harbour has 9 piers, with
depths of up to 7 m at low water. There is a
slipway and dry dock which can take ships
of up to 2 400 tons and the access to tugs is
good.

Tiksi. The harbour is located at the delta of
Lena (Figure 3.27), and is about midway on
the Northern Sea Route. It is an important
trans-shipment harbour from river boats to
seagoing vessels. The approach goes
through very shallow waters and a pilot is
absolutely necessary. There are quay
possibilities for ships with a draught of up to
6.7 m. Ships with a larger draught are
referred to anchor in the well protected bay.

Pevek. In the same manner as Tiksi this is
an important harbour on the Northern Sea
Route (Figures 3.25 and 3.27). Pevek is a
mining town with approximately 5,000
inhabitants and is the centre of icebreaking
operations in the eastern part of the route.
The harbour and the icebreakers are
normally operated by the Far East Shipping
Company. There is a quay here of 440
metres length. The draught is between 8 and
10 meters. In addition, there is an offshore
quay for seagoing tankers.

Greenland, Canada and the US

Along the coast of Greenland there are a
range of harbours with anchoring and quay
facilities for small and medium-sized ships.
Many of these berths can provide services of
a limited extent. In addition to this, there is a
swarm of protected anchor places. Detailed
information can be found in the Arctic Pilot,
Vols. 2 and 3 (NP-11 and 12).

In the Canadian Arctic the harbours are for
natural reasons seldom built with quay
installations. However, there are a range of
service centres where services are provided
to shipping. As with Greenland,there are
many well protected anchor places. Detailed
information is to be found in Arctic Pilot,
Vol.3 (NP-12). On the south coast of Alaska
there are several harbours which serve the
local fishing fleet, as well as the oil terminal
in Valdez from which the oil from Alaska is
shipped out.

In connection with the development of more
international polar shipping, one can
imagine that the harbour at Dutch Harbor
(Figure 3.26) will have a special function as
a junction for trans-shipment. At present
(2011) the harbour also has an extensive
service offer and there is an airport in the
vicinity. The harbour is also of great
importance for the fisheries in the area.

Figure 3.26
In Dutch Harbor, Alaska, efforts are concentrated on future extensive activity as a result of increased Arctic shipping.

3.5 Sailing description of the Northern Sea Route

In this chapter the waters which are directly connected to the Northern Sea Route (Northeast Passage) will be described. As regards special stretches that are not normally connected to the Northern Sea Route, reference will be made to the American and English pilot descriptions (US 183). The descriptions here will be from west to east. Usual routes are drawn in Figure 3.27. Recommended routes will also be drawn in on the Russian charts (electronic and paper). Distances of the most usual routes are to be found in Table 3.1.

Table 3.1
Distance advantage by using the Northern Sea Route (NSR) compared to the Panama Canal or Suez Canal.

Voyage:	Distance	Distance (NSR)	Diff.
Tromsø - Yokohama	12400 nm[*]	5600 nm	-55%
Tromsø - Vancouver	9100 nm[**]	5750 nm	-37%
Tromsø - Dutch Harbour	10450 nm[**]	3050 nm	-71%
Hamburg- Yokohama	11070 nm[*]	6900 nm	-38%
Hamburg- Vancouver	8740 nm[**]	6630 nm	-24%
Mo i Rana-Hong Kong	10500 nm[*]	7900 nm	-25%

[*] Via Suez
[**] Via Panama

The Northern Sea Route runs from Murmansk and Arkhangelsk in the west to Vladivostok in the east (or more precisely from the Kara Gate to the Bering Strait). It runs through the Kara Sea, Laptev Sea, the East Siberian Sea, and the Barents Sea. The navigation season normally lasts for three months, from the end of July to the end of October. However, much is being done to extend the season and shipping to Yenisei has been kept going year-round with the exception of about 3 weeks when the ice moves out of the river. The most difficult time for shipping is approximately 6 weeks in May-June when river ice moves out.

All ships that are used must be ice reinforced equivalent to the Russian classes ULA and UL (IACS PC6-PC7). This is approximately equivalent to the ice class 1A* and 1A in DNV. Russian ice classes will be described further in Chapter 5. Escorted vessels can in some instances carry a lower ice class.

Icebreakers are available upon request. In accordance with the guidelines for the voyage, foreign vessels must have an icebreaker escort and Russian ice pilot. For the western part, to125°East, the traffic is organized from Dikson. On the remaining part the service is organized from Pevek. On large parts of the route, piloting can be done in the traditional manner with a pilot on board. All seafarers are obligated to report position, weather, ice, and sea conditions twice daily, at 2400 hours and 1200 hours, UTC.

There are four possible approaches to the Kara sea. These are Proliv Yugorskiy Shar (Jugor Strait), Proliv Karskiye Vorota (Kara Gate), Proliv Matochkin Shar (Matochkin Strait) and around the northern tip of Novaya Zemlya. Choice of approach is indicated by the icebreaker service and is based on the ice situation. This shall be cleared with the station at Dikson 24 hours before one draws near Novaya Zemlya. If all the approaches are ice-free, and one does not know the ice conditions behind them, the southernmost route should be chosen. The ice conditions are usually easier there.

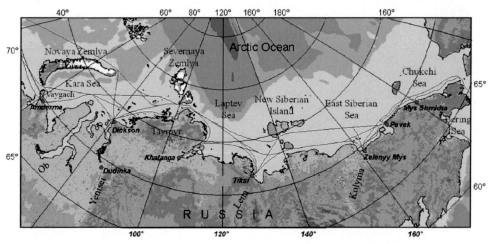

Figure 3.27
The chart shows the most important routes on the Northern Sea Route (Armstrong, 1990).

The Jugor Strait. This strait, which separates Vaygach Island from the mainland is 21 nm long. On the north side the coast is steep and provides good radar reflection. On the south side of the sound the coast is flatter and more difficult to identify. There are several shallows in the sound. Because of strong currents, navigation must be undertaken with extreme caution. Both entrances to the sound are mined, and anchoring is prohibited.

The Kara Gate. This strait, which runs between Vaygach island and Novaya Zemlya, is the most frequented entrance to the Kara sea. In the centre of the sound a separation system has been set up. Normally the ice will break up about 25 June, and pack again about 17 November. The sound is well marked with lights and Racon.

The Matochkin Strait. This strait, which divides Novaya Zemlya is narrow, but relatively easy to sail in. Depths are over 12 meters. There are quays in the sound, but at present they are all closed to all civilian shipping because of military activity. There were extensive nuclear tests in this area up to 1990.

Figure 3.28
An example of the ice situation in the winter of 2002 where sailing northward of Novaya Zemlya can be preferable to the sounds south of the island.

The north side of Novaya Zemlya. The waters and entrance to the Kara Sea are open and easy. It is important to note the danger of encountering larger icebergs. From time to time when there is a lot of ice in the Kara Sea, this route could in actual fact be the simplest (Figure 3.28).

<u>Vilkitsky Strait.</u> The strait runs between the mainland and Severnaya Zemlya. This is the northernmost point of the Northern Sea Route, and the entrance to the Laptev Sea (78°00'N, 103°00'Ø). It is also possible to get into the Laptev Sea by using Proliv Shokal`skogo between the islands in Severnaya Zemlya, but the ice conditions are usually worse there. In this area it is almost always necessary to have an icebreaker escort, since the ice has a tendency to cover down the northeast side of Taymyr Peninsula. In the case of southerly winds in the summer, it can be that the shore lead is navigable and ice-free for small vessels. All ships that pass Cape Chelyuskin on the north side of Taymyr are obligated to report their route and identifying signal.

<u>Laptev Sea.</u> From the Vilkitsky Strait the route runs eastwards towards the New Siberian Islands. Most of the sea area goes over the large Lena delta, which is very shallow (Figure 3.29).

Figure 3.29
The Laptev Sea is very shallow. The figures to the right indicate max. draught for the various routes.

On the east side of the Laptev Sea and in the <u>Sannikova Strait</u> are found the shallowest areas on the Northern Sea Route (marked with 11m in Figure 3.29). The voyage must therefore be undertaken with the utmost care. The shallowest areas which must be passed in the strait measure approximately 11.5 m. Here one shall be aware of the extensive mixing of fresh water which will lead to a deeper draught for the ship. As shown in Figure 3.29 a ship with a draught of 16m must sail approximately 80nm from the coast in order to have sufficient clearance under the keel.

<u>The East Siberian Sea.</u> Here the ice will often lie far south, especially with northerly winds. One is therefore normally referred to sail as close to land as depth allows. The largest settlement on this stretch is Pevek, which is located at approximately 170° E, and is the service harbour for the eastern part of the Northern Sea Route.

3.6 Special provisions and sailing regimes

Familiarization with current legislation in the planning and implementation of a voyage is important for all waters. This is especially valid for ice-covered waters, since in several countries there will be rules and sailing regimes connected to the conditions in a given period of time. Such information can be found in updated pilot descriptions and publications which are available in the administrations of the various countries.

3.6.1 Russia
In 1991 the Northern Sea Route was officially opened for international traffic after having been closed in principle since the Russian revolution in 1917. This also entailed that a set of rules in the English language was issued (Figure 3.30). The area of application of this is the same as that which is defined as the Northern Sea Route – that is to say, the areas between Novaya Zemlya and the Bering Strait. It is worth

noting that Russia considers the area as an inner sea route, and thereby not as an international passage in the sense of the law of the sea.

The fundamental principle for the rules is that it is intended that ships sailing the route shall have the support of powerful icebreakers. This means that the requirement for ice reinforcement can be somewhat reduced. The main features of the rules are as follows:

- Minimum ice class equivalent to IACS PC7 (1A in DNV).
- Hull without bulbous bow.
- Extra bunkers capacity.
- Own ice pilot.
- Inspection of the ship.

Figure 3.30
Rules for voyages along the Siberian coast.

There is still little practical experience with fees for through-sailing, but tariffs have been drawn up which are based on the ship's tonnage, and the number of any passengers. This fee guarantees support of icebreakers if it is necessary (Figure 3.31). Today (2011) it is estimated that the icebreaker fee will be in the region of USD 50,000 /day. The

icebreaker fleet consists now of 6 operational nuclear icebreakers and 4-6 Arctic diesel icebreakers. In addition, a few "smaller" icebreakers will be available, which are usually used in areas with only one-year ice (White Sea, Okhotsk Sea and The Gulf of Finland). Some of these ships will be described in Chapter 5.

Figure 3.31
A bulk carrier is escorted in difficult ice conditions through the Kara Sea by two nuclear icebreakers.

3.6.2 Canada

In the Canadian Arctic there are in principle two different regimes for regulating navigation in ice. The regimes are called CASPPR and CAC, and both are based on that ships should sail unassisted by icebreakers to their destination. This is fundamentally different from the regimes in Russia and Gulf of Bothnia.

CASPPR (Canadian Arctic Shipping Pollution Prevention Regulations):
This is a relatively rigid system where the Arctic has been divided into special zones which limit the traffic in relation to a ship's ice class and the time of the year (Figure 3.32). The principle is that a given ship that carries a given ice class will be granted permission to sail into a zone in a fixed defined period of time. The disadvantages of this are clearly that the annual variations in ice conditions are disregarded, and consequently the safety margins must be relatively conservative.

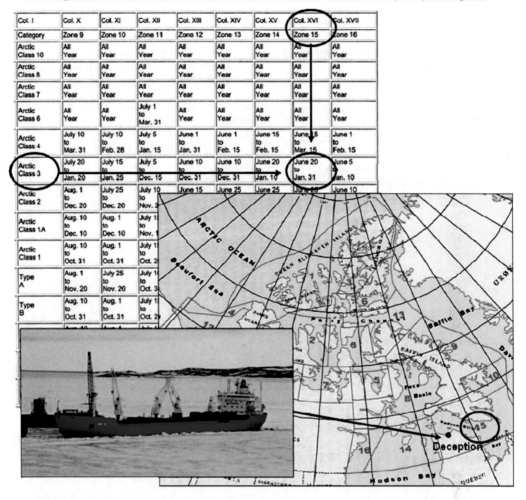

Col. I	Col. X	Col. XI	Col. XII	Col. XIII	Col. XIV	Col. XV	Col. XVI	Col. XVII
Category	Zone 9	Zone 10	Zone 11	Zone 12	Zone 13	Zone 14	Zone 15	Zone 16
Arctic Class 10	All Year	All Year	All Year	All Year	All Year	All Year	All Year	All Year
Arctic Class 8	All Year	All Year	All Year	All Year	All Year	All Year	All Year	All Year
Arctic Class 7	All Year	All Year	All Year	All Year	All Year	All Year	All Year	All Year
Arctic Class 6	All Year	All Year	July 1 to Mar. 31	All Year	All Year	All Year	All Year	All Year
Arctic Class 4	July 10 to Mar. 31	July 10 to Feb. 28	July 5 to Jan. 15	June 1 to Jan. 31	June 1 to Feb. 15	June 15 to Feb. 15	June 15 to Mar. 15	June 1 to Feb. 15
Arctic Class 3	July 20 to Jan. 20	July 15 to Jan. 25	July 5 to Dec. 15	June 10 to Dec. 31	June 10 to Dec. 31	June 20 to Jan. 10	June 20 to Jan. 31	June 5 to Jan. 10
Arctic Class 2	Aug. 1 to Dec. 20	July 25 to Dec. 20	July 10 to Nov. 2	June 15	June 25	June 25	June 25	June 10
Arctic Class 1A	Aug. 10 to Dec. 10	Aug. 1 to Dec. 10	July 15 to Nov. 1					
Arctic Class 1	Aug. 10 to Oct. 31	Aug. 1 to Oct. 31	July 15 to Oct. 2					
Type A	Aug. 1 to Nov. 20	July 25 to Nov. 20	July 10 to Oct. 3					
Type B	Aug. 10 to Oct. 31	Aug. 1 to Oct. 31	July 15 to Oct. 2					

Figure 3.32
According to the zone / date system in CASPPR, the MV Arctic cannot sail to Deception Bay in May. The table above shows only zones 9 – 16.

In Figure 3.32 a concrete example is shown with the bulk carrier *Arctic*. The ship is reinforced to ice class Class-3 according to Canadian definitions. If the ship's destination is Deception Bay in the Hudson Strait to load, the period in which this can be done in Zone No. 15 must be checked. From the table it will be seen that sailing with this ship in this zone can be done between 20 June and 31 January. If one has a ship which

carries an ice class in DNV / IACS, the equivalent tables must be consulted to find what this means in relation to the Canadian definitions in the table. As we will see, Arctic-10 class is required to navigate the whole year in all areas. Canada had plans to build such a ship which would be equipped with 100,000 hp, but the project was shelved for financial reasons.

Because of the relatively rigid structure in CASPPR, in recent years a new and more flexible system has been developed, which will be preferable in the future. This is called CAC (Canadian Arctic Class).

CAC introduces a completely new train of thought for regulation of ice navigation, and here also there has been a thorough review of the rules for ice reinforcement of ships. Icebreakers were classified as CAC 1 – 4, and other ice reinforced ships as CAC a – e ("type ships"). The principle for sailing permission shall then be granted by a calculation in which is included the ship's ice class (A), ice conditions / environment (B), and operational conditions (C). The calculation for permission shall therefore be given by that:

$$A + B + C > 0$$

The codes which are included in A, B and C will be found in separate descriptions (CCG, 1992). The figure which emerges from the calculation is called "Ice number / Ice numerals".

Example:
Take the case that the conditions are relatively favourable and that the MV *Arctic* will attempt to said to Deception Bay in May (before the CASPPR regime grants permission). A thorough study of the ice chart must be done first (ref. Figure 3.36) and the route planned accordingly. Based on this and the ship's ice class, the ship may seek permission from NORDREG Canada which is the authority that grants permission. The application which is forwarded to NORDREG could like the telex in Figure 3.33. For more detailed information, see the website of the Canadian Coast Guard (http://www.ccg-gcc.gc.ca/eng/MCTS/Vtr_Arctic_Canada).

From the application can be seen that the calculated ice numerals are all positive figures (> 0). The result of the application was that the *Arctic* received permission four hours later.

In addition to the requirements made by the CASPPR and CAC regime, it can be experienced that the local authorities in Nunavut can make requirements regarding

paying attention to the needs of the local population, i.e. the ice on some fjords can be an important arterial road for parts of the year. Breaking up the ice can therefore be prohibited before a given date.

```
RCR.RR              Page  1     UTC Time: 99-05-26  19:02:13

TO:ECAREG CANADA
ATTN: NORDREG CANADA

ARCTIC
VCLM 7517507
ARCTIC CLASS 3
26/05/99 1805UTC
POSITION: 60 00N  059 00W
CO: 310(T)  SP: 14.5KTS
ICE: BERGY WATERS
DESTINATION: DECEPTION BAY
INTENDED ROUTE FROM PRESENT POSITION
TO (A) 61 15N  062 00W
TO (B) 61 15N  064 55W
TO (C) 62 28N  070 42W
TO (D) 62 30N  074 26W
TO DECEPTION BAY

ICE REGIMES A TO B:
9/10 TOTAL, 5/10 TFY, 4/10 MFY, 1/10 THIN FY

B TO C:
3/10 TOTAL, 1/10 TFY, 2/10 MFY

C TO D:
VARYING FROM 3/10 TOTAL, 1/10 TFY, 2/10 MFY
TO 9/10 TOTAL, 3/10 TFY, 6/10 MFY

D TO DECEPTION BAY:
6/10 TOTAL, 4/10 MFY, 2/10 THIN FY

ICE NUMERALS:  A-B: 20
               B-C: 6
               C-D: FROM 6 TO 18
               D-DECEPTION BAY 12

SOURCES OF ICE INFO: LABRADOR COAST ICE ANALYSIS CHART OF 25/05/99
                     ICE IMAGERY OF HUDSON STRAIT 25/05/99 FROM
                     NOAA SATELLITE

ICE NAVIGATORS: D.MILLAR (MASTER)
                D.KRUGER (RELIEF MASTER)


REGARDS: MASTER

SENT VIA LABRADOR CGR  (2514/2118) 1850UTC 26/05/99
```

Figure 3.33
Application for entrance to Deception earlier than CASPPR allows.

The description above concerns mainly voyages in the Arctic regions of Canada. For more information on ice conditions in the St. Lawrence Gulf and on the Lakes, the Coast Guard web pages (www.ccg-gcc.gc.ca) is recommended, as well as the pages of the American Coast Guard:
http://www.uscg-iip.org/
http://www.navcen.uscg.gov/
http://uscg.mil/

3.6.3 Sweden and Finland

The same as in Russia, the regime in Gulf of Bothnia is based on massive support of relatively powerful icebreakers (Figure 3.34).

The criteria for whether a ship will receive permission to sail to its destination in Gulf of Bothnia will be laid down by the authorities based on existing ice conditions. As in Canada, the starting point will be standardized criteria for the ship's ice reinforcement (ice class). The ice classes, also known as fee-classes, which Finland and Sweden drew up in this connection, have also formed the basis for the technical descriptions of the ice classes in the classification companies.

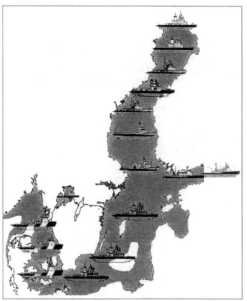

Figure 3.34
The chart shown how Swedish and Finnish icebreakers can be distributed in Gulf of Bothnia in a normal ice winter.

The philosophy behind the Finnish and Swedish fee classes was that a powerful ship with a high ice class should pay a smaller pilot fee. In this manner, one has encouraged shipowners to build better and better ships

for winter traffic in Gulf of Bothnia. In addition to the ice class there will be a requirement connected to that the ships shall have a minimum tonnage. This is because a heavier ship will normally be better able to navigate in difficult ice conditions. How this is represented in a typical winter is shown in Table 3.2. How notification regarding restrictions can look is shown in Figure 3.35.

Information on restrictions can be obtained from coastal radio stations in the area or by checking the Maritime Authorities' web pages: http://www.sjofartsverket.se/ www.baltice.org/

As mentioned, the services will be based on extensive icebreaker support. Ships that wish to enter the area should therefore check that they satisfy the requirements laid down, i.e. necessary arrangements for towing by icebreakers (ref. also Chap. 6).

Table 3.2
Typical minimum requirements for ice class, tonnage and time for sailing in Gulf of Bothnia in a typical ice winter.

Min. ice class / min.dwt	Bothnian Bay, date	Bothnian Sea, date
II /1 300	1/12	1/1
II /2 000 1C /1 300	15/12	15/1
1B /2 000	1/1	1/2
1A /3 000	15/1	15/2
1A /4 000	31/1	–
1A /3 000	10/4	–
1B /2 000	10/5	1/4
II /2 000 1C /1 300	15/5	15/4

Traffic restrictions: Traffic restrictions are imposed to improve the efficiency of vessel traffic. **Icebreaker assistance will only be given to vessels, wich meet the requirements set out in the traffic restrictions.**
Present restrictions: Karlsborg - Lulea: Minimum 2000 dwt and at least iceclass IB **Haraholmen - Skelleftehamn:** Minimum 2000 dwt and at least iceclass IC **Holmsund - Ornskoldsvik:** Minimum 2000 dwt and at least iceclass II **Upper river Angermanalven:** Minimum 2000 dwt and at least iceclass II **Lake Malaren and Lake Vanern/Trollhatte Kanal:** Minimum 1300 dwt and at least iceclass IC or minimum 2000 dwt and at least iceclass II

Figure 3.35
Notification of the minimum requirements
for ice class and tonnage in various areas
(13 December 2010).

3.6.4 Svalbard

Previously there has been little restriction regarding sailing at Svalbard, apart from circumstances concerning expedition activity and landing. In 2008, however, the Harbour and Waters Act also became applicable to Svalbard. This will mean that ships must conduct themselves in relation to sailing in territorial waters on the same lines as that applicable to the mainland. Chapter 8 of the Regulation relating to maritime traffic describes amongst other things special new obligations in the Aksel Strait through which is the entrance to Svea. The new regime also entails a reporting obligation and various navigational considerations. Updated information regarding this can be found on www.kystverket.no or in Den Norske Los, Vol. 7 (new edition 2011). In the revision of the Pilotage Act (2010) new rules are included in relation to the requirement of having a"Polar pilot" in the waters surrounding Svalbard.

3.6.5 Antarctic

Historically, there has been a minimum of restrictions in relation to the international shipping activity in the waters surrounding Antarctic. The Antarctic Treaty does not deal with this type of problem, and it has been up to the operators themselves to safeguard their operations. Thus, the International Association of Antarctic Tour Operators (IAATO) has played a serious role by recommending ice reinforcement on ships in relation to time and area – much the same as is found in the zone/date system in Canada. These recommendations are shown in Figure 3.36 (ref. also www.iaato.org). IMO has also included the area in what is called "special area" in the MARPOL Convention, which means a stricter regime in relation to emissions. In the rules from IMO it is also stated that ships are prohibited from using heavy fuel oil in these waters.

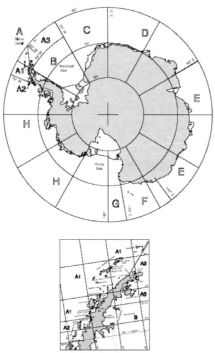

Polar Class	PC 7	PC 6	PC 5	PC 4- 1
Similar other classes	RRS L1-3; ABS D0-1A; DNV C-1 A; GL E3	RRS L4-6, ABS AA/A0, DNV 1A*; GL E4	RRS L7-9; ABS A1; DNV 1A*F; GL Arc 2-3	RRS LU6-9; ABS A2-5; DNV Polar 10-30; GL Arc 1-4
Zone	Navigational Period			
A1	1.12.-20.02.	1.12.-20.2.	all year	all year
A2	1.12.-20.02. only seawards off islands	1.12.-20.2.	all year	all year
A3	1.1.-20.2., if ice coverage < 5/10	1.12.-20.2.	1.10.-30.4. W-wards from 40˚ W; all year E-wards from 40˚ W	all year
B	never	1.1.-20.2. except in SW quadrant	all year in NE and SE quadrant; 1.1.-20.2. in NW + SW quadrant	all year
C	1.1.-20.2.	1.12.-20.2.	all year N-wards from 70˚ S; 1.10.-15.3. S-wards from 70˚ S	all year
D	1.12.-20.2. N-wards of line Erskine Iceport - Amundsen Bay		all year N-wards of line Erskine- Amundsen; 1.10.-30.4. everywhere	all year
E	1.1.-20.2. N-wards of West+ Shackleton ice shelf	1.12.-20.2. N-wards of West+ Shackleton ice shelf	all year N-wards of West +Shackleton ice shelf; 1.12.-30.4. everywhere	all year
F	never if ice coverage >5/10	1.1.-20.2. N-wards of line Balleny	all year N-wards of line Balleny; 1.1.-20.2. everywhere	all year
G	1.1.-20.2.	1.12.-20.2.	all year	all year
H	never if ice coverage >5/10	1.12.-20.2 N-wards of line Adelaide Isld-75˚S/160˚W	all year N-wards of line Adelaide Isld. - 75˚S/160˚W; 1.10.-30.4. everywhere	all year

Figure 3.36
Recommended ice reinforcement of ships
that are to sail in the Antarctic.

3.7 Questions from Chapter 3

1)
Describe special conditions which make navigation (positioning) difficult in many polar waters.

2)
Why must navigation in front of glacier fronts be undertaken with the greatest care?

3)
Where will you find information on sailing conditions along the Siberian coast?

4)
Give examples of operational conditions which can contribute to better safety for voyages in remote regions.

5)
Will a satellite navigation system function at higher latitude than 80° N/S?

6)
Which communications system would you use to telephone from the Arctic Ocean?

7)
Are there differential support systems for GPS in the areas around Svalbard?

8)
What mistakes can you expect to occur on a conventional gyrocompass if you at 79°N sail northwards at 16 knots?

9)
How can you make a celestial observation with a sextant when the horizon is disturbed by ice?

10)
What distance advantage can be achieved by using the Northern Sea Route between Tromsø and Yokohama, in relation to sailing through the Suez Canal?

11)
What ice class will normally be required to sail the Northern Sea Route?

12)
Why is it not practically possible to said the straits which divide Novaya Zemlya?

13)
What other challenges than ice conditions can be expected when you are to sail the Northern Sea Route?

14)
What is the main principle of CASPPR?

15)
Which period in time could you sail to the northern tip of Baffin Island with a ship carrying Arctic Class 3 according to the Canadian rules?

16)
Which limitations / requirements are imposed on ships that are to sail to Gulf of Bothnia in winter?

17)
Give an example of where you will find guidelines for sailing in the Antarctic.

4 Ice mechanics

4.1 Ice crystal and ice formation

Ice formation will normally take place when the water emits heat after it is cooled down to freezing point. The location of the freezing point is dependent on several variables which will be described later on.

Theoretically, the ice can absorb several different molecular bindings, but in nature only hexagonal bindings are found, as shown in Figure 4.1.

Chemically, the ice consists of oxygen and hydrogen which is bound together in a tetrahedral form (Figure 4.2). The four oxygen atoms will be located at a mutual angle of approximately 109°, with the hydrogen atom in the middle. The structure arises after the density of the water makes a marked leap at the transition from water to ice (Figure 4.3).

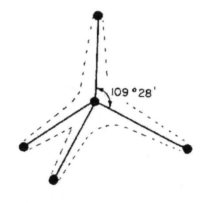

Figure 4.2
Tetrahedral form on oxygen atom in ice.

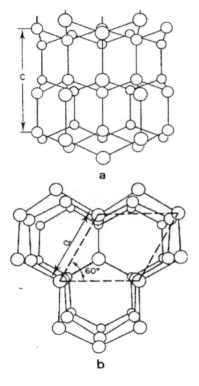

Figure 4.1
Molecular structure of ice. The lines indicate the connection between the oxygen atoms (hydrogen is not shown). Figure a) is shown perpendicularly on the c-axis and Figure b) along this (Michel, 1978).

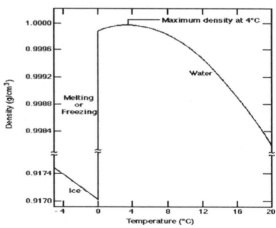

Figure 4.3
The relationship between density and temperature in fresh water. The ice will float because of its marked lower density than water.

In seawater there will also be a range of different salts which will be of great significance to the crystal structure in the freezing process. This will cause a different and more complex context between density and temperature than that shown in Figure 4.3. In seawater the total amount of salts represents a given number of thousand parts (per mille) of the mass unit. This is called salinity. As an average in the world's oceans it is usual to regard the salinity as 34.5 per mille. Because of the salt content of seawater, the freezing point will fall under 0°C. Figure 4.4 shows the relationship between salinity and temperature, and in general one can say that cold salt water is heavier than warm brackish water and ocean water, which is more saline, has a lower temperature of maximum density and a lower freezing temperature than fresh water.

increases, frazil ice will be formed (brash). As the freezing process continues, the first layer of ice will be formed.

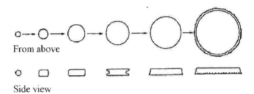

Figure 4.5
Growth of disk-shaped ice crystals.

The structure of the first ice is very dependent on the salinity. Clean water is crystallised, while the salts and brine are collected in pockets which are connected to fine channels as shown in Figure 4.6.

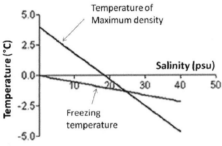

Figure 4.4
Relationship between salinity and temperature of water to illustrate how salinity influences the temperature of maximum water density and freezing temperature.

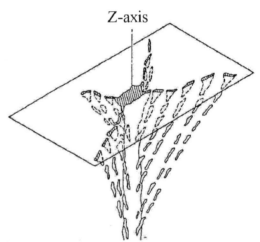

Figure 4.6
Sketch of salt channels in sea ice. The Z axis represents the vertical.

The ice formation itself will take place in the same manner in seawater as in fresh water. The first ice crystals to be formed will grow to small slivers which will be maximum 2 - 3mm in diameter (Figure 4.5).

Because of the influence of wind and waves, however, the ice slivers will mix with a deeper water level and we will have super-cooled water down to a given depth. Gradually as the concentration of ice slivers

If the ice is cooled down further, the amount of brine in the channels will decrease and the salt will be crystallised to solid form (Table.4.1). At temperatures above -15°C the salt channels will act as draining of brine. The main channel which in the figure is indicated as the Z-axis can then be 1 cm in size. The horizontal distance between the main channels is observed to be a couple of metres in normal sea ice. When the freeboard of the ice increases the gravitation will drain the brine and the solid salt downwards through the ice. We can illustrate the salinity in the salinity profile as shown in Figure 4.7. Figure 4.7 also shows the growth of the ice.

Table 4.1
Weight distribution in per cent between ice, brine and salt from a sample of typical sea ice.

Temperature [°C]	Ice	Brine	Salt (solid)
- 1.9	–	0	0
- 10	76.8	22.8	0.42
- 30	91.7	3.95	4.3

more parameters, which are connected to salinity in ice and underlying water. Growth will then be:

$$\frac{dh}{dt} = \frac{1}{g\rho'L} \cdot \left(\frac{1}{h/K_i + h_s/K_s} \right) \frac{dS}{dt}$$

ρ' = density of solid ice
g = gravitational acceleration
K_i = thermal conductivity co-efficient of the ice
K_s = thermal conductivity co-efficient on snow cover
L = latent heat in the ice
h = ice thickness at the time , t
h_s = snow depth
S = number of "day degrees" of frost

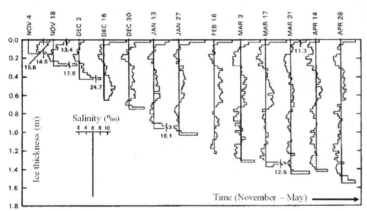

Figure 4.7
Vertical salinity profiles in sea ice and the growth of the ice. The intervals between the profiles are 2 weeks. Data basis is from the Eclipse Sound in Canada, winter 1977-78.

Many attempts have been made to calculate the rate of growth of the ice, but this is far more complicated than one would believe initially. It is dependent on temperature, snow cover, density, ability to conduct heat, ice thickness, salinity, etc. In order to illustrate this problem we use an equation which describes the growth of the ice (dh/dt) in fresh water. The equation is based on the theory of thermology. For sea ice the growth equation will be the same but will contain

In addition to such theoretical calculations it is usual to illustrate the growth of ice in an area based on statistical data from annual measurements (Figure 4.7).

In Figure 4.8 we see that the temperature of the ice rises as a function of depth. If we had performed an equivalent analysis of icebergs which are far thicker, we would have noticed that the temperature would stabilise a few metres below the surface. This temperature would lie quite close to the average temperature for the area where the iceberg was produced. Researchers can use this technique to determine the origination of icebergs in the Northern Barents Sea. Icebergs which originate in Franz Josefs Land will e.g. have a core temperature which lies a few degrees lower than the icebergs from Svalbard.

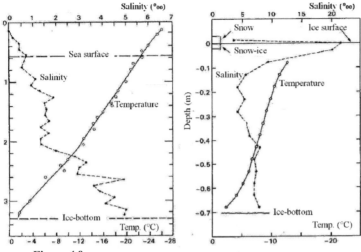

Figure 4.8
Temperature and salinity profile in ice. To the left in a packed multi-year floe, to the right in first- year ice from the Barents Sea.

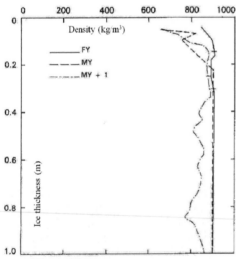

Figure 4.9
Density profiles in first-year ice (FY), multi-year ice (MY), floes which the summer before the measurement were identified as multi-year ice (MY+1).

4.2 Physical and mechanical properties of ice

Seen from a shipping point of view, perhaps the most interesting property of ice is its strength. However, the strength is dependent on several parameters such as temperature, density, porousness, etc. These are measurements that are bound together. It is also usual to divide the strength factor up into tensile / compressive strength and loading capacity. The density can vary somewhat as a function of thickness and age. This is indicated in Figure 4.10. If the ice is pore-free and fresh, the density can be described as the equation below:

$$\rho = 0.9168(1 - 1.53 \times 10^{-4}T) \quad [g/cm^3]$$

where:

ρ = density
T = temperature in °C

The effects of temperature are connected to the expansion co-efficient, but this is less than the effects of porousness. Normally, sea ice will be weaker than fresh ice, but at very low temperatures the case will be the opposite. At -5°C sea ice will have a compressive strength of approximately 3.5MPa. The tensile strength will be approximately half of the compressive strength. The compressive strength of the ice is of great importance when we look at the capacity of an icebreaker or the load on other ships which move through ice-covered areas. The granule size of the ice is also significant and from Figure 4.10 we can see that the larger the granules, the weaker the ice will be.

70

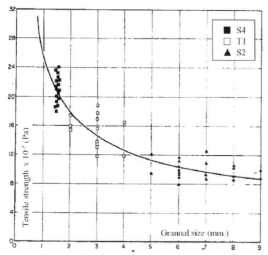

Figure 4.10
The tensile strength of the ice as a function of the granule size. S4, S2 and T1 are different "ice classes" depending on the size of the granules (Michel, 1978).

As if we have not mentioned enough variables for the strength of the ice, it is necessary to look at the elasticity of the ice (elasticity modulus). The elasticity is dependent on the time interval it is exposed to forces. Quick powerful impulse gives low elasticity and reduced resistance to deformation.

During many arctic operations it is relevant to use the ice as a loading / unloading place. In many arctic areas it is also usual to use the ice for roads and landing strips for aircraft. With such operations it is important to know the **load-bearing capacity** in order to ensure the safety of the crew and equipment.

The load-bearing capacity and deformation are derived from the theory we know from mechanics. We look at an ice girder which is supported at both ends and exposed to load from the top (see Figure 4.11). This is also theory which is important in the calculation of ice loads on ships and constructions. On a girder as shown in Figure 4.11 compressive forces will occur on the top and

tensile forces at the bottom. If we compare this with the fact that the ice is weaker for tensile, we will see that the weak point is at the bottom.

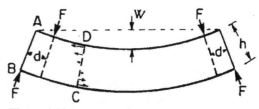

Figure 4.11
The load-bearing capacity of ice can be regarded as an ice girder similar to what we know from mechanics. Deflection, W, is dependent on the size of the force (F) and its distance from the supporting points. The pressure and tensile forces are shown in the cross-section C-D.

Many experiments have been made with the load on ice girders in order to describe the properties of the ice. Figure 4.12 shows deflection on the ice girder which is loaded at four points at T= -3°C, as a function of the time. Girder A is loaded with 66 kg, and we see that a rapid deflection occurs followed by a creeping change over a long time. At time, t, it is indicated that the weight is removed and we see a weak elastic retraction. If we look at the curve for girder B, which is loaded with 99 kg, we see that it does not manage the transition from rapid deflection to slow creeping change. The girder will have cracks underneath and break. Figure 4.13 shows what happens with the ice girders if they are loaded gradually. Here are the three girders which are loaded in the same way as above. The temperature here is -10°C. A is loaded with 99 kg +10, 11, and 12 kg, gradually to breaking point. B is loaded with 99 kg +10, +11 kg. C is loaded with 109 kg + 11 kg. The ends of the curves indicate breaks. We see that with gradual load we can achieve the greatest deflection and delayed break. All the girders break without acceleration of creeping.

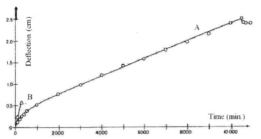

Figure 4.12
Deflection on an ice girder as a function of the period of load (temp. = -3 °C).

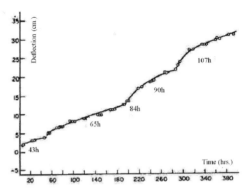

Figure 4.14
Deflection on ice which is gradually weighted with the load from a basin which was filled with water. The diameter of the basin was 7 m.

As a rule of thumb one can say that the ice should not be loaded to a greater deflection than the freeboard of the ice. The reason for this is that if a crack occurred, the ice would be flooded with water which must then be regarded as additional weight. Purely practically it could also lead to greater difficulties if the ice was flooded.

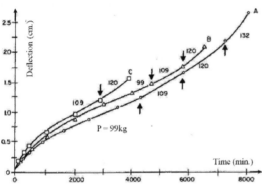

Figure 4.13
Deflection on ice girders as a function of time. The curves A, B, and C indicate various loads, specified in the text above (Nadreu, 1978).

We have now regarded the ice as a girder. In reality, we must naturally relate to an ice cover, but in principle the load response is very similar to the one we have studied for the girder. Figure 4.14 shows this from a field experiment which was undertaken on 1.33 m thick sea ice. A basin measuring 7 m in diameter was put on the ice and filled with water by stages at fixed intervals over a period of 380 hours. The measurements of the deflection (W) show the same tendencies as in the case of the girder – first a rapid change after the load, which after a short period of time goes over to an even creep.

Dynamic loading.
As will be seen from the previous pages, the ice can have different properties based on the time the load lasts. We are talking about elastic or plastic response. In the case of quick loading the ice will have elastic properties while with gradual or consistent loading it will have creeping properties (plastic).

This is of significance if we drive a vehicle or try to land an aircraft on the ice. Experiments have shown that at speeds of more than 3 m/s (10 km/h) the ice shows an elastic response. From 3 m/s and down to 3 cm/s (0.1 km/h) we will have a progressive plastic response. Below this speed we expect that the movement will give a total plastic response. Since most movable loads are in the elastic response area it will be of interest to find the load-bearing capacity for an elastic ice-cover. For this, very simple empirical formulas have been developed in order to indicate how many kilograms, P, one can move on the ice with a given thickness, h.

$$P = A \ x \ h$$

(P in kilograms, h in metres, A is not designated)

A is a constant, given by ice type and characteristics. $A = 350 \ x \ 10^2$ is used a lot for landing of aircraft on fresh water ice in Canada. On sea ice this figure will be considerably lower.

Example:
The ice on an inland lake is 50 cm thick. How heavy an aircraft can land?

Solution:

$P = 350 * 10^2 * 0.5 = \underline{17,500 \ kg.}$

An aircraft weighing **17.5** tons can land on this lake.

NB !
It is worth adding a couple of remarks to this calculation:

i)   At high daily average temperatures, above -1°C for fresh water and -2°C for sea, the load should be reduced by 10% per day. The operation should be discontinued after 4 days.

ii)   In the event of movement on the ice, a critical speed can be reached which is in resonance with the relationship between the extent of the ice and the depth of the water. The wave which is created under the ice can overstress the ice.

4.3 Static and dynamic ice loads on marine constructions

For all constructions that could come into contact with ice it is important to dimension for the loads the ice can impose on the construction. The problem is relevant for ships, platforms, dams, quays, etc. Concerning the force from icebergs and ice islands, these are so extensive that it would be unrealistic to dimension with regard to them. In such cases one is faced with an operational problem which will be described in Chapter 6.

Here, four categories of ice loads on marine constructions will be described. Concerning icing on constructions, this will be described in Chapter 6.

1)
Static pressure from an ice cover which is under expansion. The expansion is a result of rising ambient temperatures or strong sunlight. Typical problem areas are dams, dikes or constructions in enclosed waters.

2)
Load from a drifting ice cover (drift ice, pressure ridges and icebergs). In the Arctic it is very important to know this load which will be of importance to ships, platforms, quays, etc. The problem will be more or less equivalent in rivers and delta areas when the ice begins to break up in spring and summer.

3)
Load from an accumulated ice mass / slush around a construction .

4)
Vertical forces from ice which is moved by Tidewater, or accumulated by horizontal movement.

Static pressure.

Static pressure caused by thermal expansion has created a lot of worry in construction of dams in cold regions. In the past it was dimensioned to tolerate that the ice was crushed against the construction. In recent years the plastic properties of the ice under pressure have been taken into consideration. This means that one could reduce the "design pressure". In Canada it is usual to dimension dams etc. for a horizontal ice pressure of 15 - 22 tons/m². The pressure that occurs will be very dependent on the angle between the ice and the land on the surrounding beaches (Figure 4.15). If the surrounding land is slightly sloping the pressure will typically be 5 - 7 tons/m². In the case of steep slopes, the pressure is measured to 26 tons/m² in periods with extreme expansion.

The thermal expansion can be due to increasing air temperature. In some cases, sunlight can also be a significant heating factor, especially at "low" latitudes.

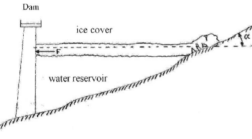

Figure 4.15
F indicates the force against a dam from thermal expansion of the ice cover. The slope of the land (α) is decisive for the load.

The thermal ice pressure can be described by the following equations:

$$\dot{\varepsilon} = \alpha \cdot \theta$$

where

$$\alpha = (54 + 0.18\,\theta)\,10^{-6}\ [°C^{-1}]$$

ε = time rate for the change in the thermal pressure [s^{-1}]
θ = rate of temperature change [°C/s]
α = thermal expansion co-efficient

Example:
If we take an average value of $\alpha = 52 * 10^{-6}$, per °C, an ice cover of 1 km can expand with 104 cm with a temperature rise of 20°C.

It is obvious that beaches and constructions which are to absorb this pressure will be exposed to large loads.

Since the ice has large thermal inertia the expansion process is something that will take place over a relatively long period of time. Rapid temperature differences in the air will have little effect. If in addition there is snow cover on the ice, this will insulate the ice from temperature variations. This condition is shown in Figure 4.16 where:

h_s = snow thickness of the ice (0 cm, 22 cm)
θ_i = ice temperature [°C].
θ_A = air temperature [°C].

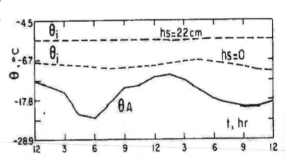

Figure 4.16
The figure shows that the snow cover will insulate the ice from the ambient temperature. We can also see the inertia in the ice temperature.

The thermal expansion will occur mainly in the upper layer of the ice and one can therefore assume that the ice pressure will be dependent on the thickness of the ice if the thickness is over 30 cm. Based on a series of pressure tests of ice, one has arrived at the conditions which are shown in Figure 4.17, where one can find the ice pressure in tons/m as a function of the initial temperature and the rise in temperature.

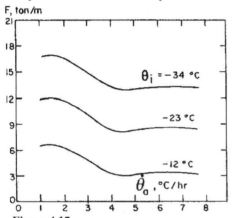

Figure 4.17
The pressure from an ice cover dependent on linear temperature rise rate in the ambient air (Michel, 1978).

Concerning the influence of sunlight at the various latitudes it is possible to make theoretical comments, but since these do not apply if there is snow on the ice, they can be regarded as hardly relevant in a nautical context.

Protection against static pressure from thermal expansion can be an air-bubble installation which will break up and melt the edge of the ice. Flexible fenders which absorb the expansion from the ice will also be a possibility – the fenders must naturally then be made of material which retains its flexibility at low temperatures.

Dynamic pressure from drift ice.
The following parameter will be of significance for calculation of the kinetic energy and the load which a construction is exposed to by contact with drift ice (Figure 4.18):

- Speed of the ice.
- Thickness.
- Temperature.

The speed in rivers will be determined by the current. At sea, the speed will be determined by both the current and wind speed. Estimates of drift speed for different types of ice under different wind and current conditions are found in literature. Knowledge of this is important in strategic planning of arctic operations in the marginal ice zone.

The thickness is a necessary measurement for knowing the mass which is moving. (Kinetic energy = $1/2 * m * V^2$). For exact calculation of the kinetic energy of the ice floe one must consider the hydrodynamic additional mass, i.e. that the calculated mass increases by 50% in relation to the real mass of the ice.

The temperature is a variable for the strength of the ice, σ. The arctic sea ice strength will increase with decreasing temperature.

Figure 4.18
This drilling platform which is anchored to the bottom will be exposed to large dynamic ice forces. We can see that the ice is packing on the one side. On the lee side, channels will open. Dynamic analysis is necessary.

In the case of collision with the ice, the ice will creep, be crushed or go through a plastic collapse a distance away from the loaded construction. Then a pressure ridge could develop.

We can do a simplified analysis by looking at the ice area, A, which collides with the construction:

$A = D * h$

(h = ice thickness, D = exposed diameter)

Load, P, is then given as:

$P = A * \sigma_c = (A * A_0)^{1/2} * \sigma_0$

(σ_c is stress by crushing)

For a normalised area, $A_0 = 100m^2$, an average value of $\sigma_0 = 0.92$ MPa, can be used.

If we consider a collision with a drifting ice floe, the contact surface (the area) will increase gradually as the floe is split, settles down or drifts past. This will lead to increasing load during the period after the collision. Figure 4.19 shows a typical curve of the load as a function of time. The rapidly fluctuating load is due to the collapse of ice in the contact surface. The frequency of the fluctuations is typically 1 - 2 Hz. In the case of such loads it is important to be aware that the natural frequency of the construction does not vibrate and strengthen the load. The average load is between 60 - 75% of the maximum load.

In order to reduce the dynamic load imposed by the drift ice, a conical ice belt is often constructed, which makes the ice break up or down instead of being crushed against the construction.

If we go further and imagine an iceberg instead of drift ice, the kinetic energy of this will be so large that it will do extensive damage to any marine construction.

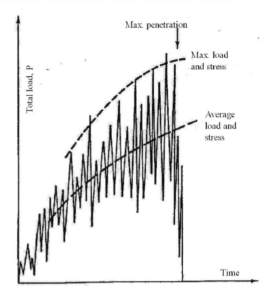

Figure 4.19
The load (P) on a structure varies with time. If the curve is integrated we will be able to calculate the total energy which the ice floe has imposed.

Load from accumulated ice mass.
Accumulated ice masses are often a problem in rivers when the outgoing ice meets constrictions or other obstructions. In the sea this can occur when relatively thin drift ice cannot get past a construction. The ice masses will then impose a horizontal force on the construction as indicated in Figure 4.20. In harbours with intensive icebreaking an ice mass could also accumulate which can be considered as "slush". In the case of such accumulations of ice the friction forces of the ice could be considerable, without having large peak loads.

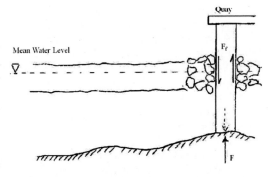

Figur 4.21
Constructions which are anchored to the bottom can be exposed to extensive friction forces if the water level changes because of tides or varying water level in a river. This does extensive damage to quay installations.

It is an important that the foundation for the cylinder is dimensioned to tolerate the vertical forces. To reduce this force, the diameter or friction can be reduced. The latter can be done by bubble installations or making the ice belt especially slippery.

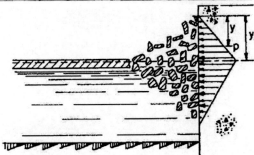

Figure 4.20
The icebreaker Tulpar clears ice along a quay in the Caspian Sea. Accumulated ice ("slush") against a vertical surface can create large horizontal and vertical forces

Vertical ice load.
An ice cover which is exposed to tidewater will move in a vertical direction and could therefore impose forces on constructions which in some cases have proved to be fatal. It is mainly the friction forces (F_f) between the construction and the ice which are of significance. If we take a frozen cylinder which is anchored to the bottom (Figure 4.21) the forces in a vertical direction will be given by the area of the contact surface (diameter * π * ice thickness) and friction co-efficient between the ice and the construction.

4.4 Classification of ice

The drift ice in the Arctic consists of three main types with differing origination. This is sea ice, river ice and glacial ice. Along the Siberian coast there are large amounts of river ice which originate from the large Russian rivers. The glacial ice originates from the calving glaciers on the northernmost island groups (Figure 4.22).

The freezing point of sea water depends on the salinity and increases with it. The salt also increases the density. In fresh water the density is largest at +4°C, while in seawater with more than $25^0/_{00}$ salt, it is largest at freezing point. The result is that cooled water will sink and start a vertical circulation. The whole of the upper water column must therefore be cooled before the freezing process starts. In quiet water, thin shell ice could form on the surface. In wind this will be mixed with the upper water layer, and brash will be formed (Figure 4.23). If one is in drift ice and where there are areas with brash ice, the colour of the brash ice can indicate increasing ice pressure. The brash ice then has a tendency to be white. Such brash ice can in some cases be reasonably thick and relatively heavy for a ship to sail in. The brash ice will normally float up and freeze solid in the thin surface ice if such has been formed first. Wind and sea will break this up in small pieces which pack against one another, and pancake ice is formed (Figure. 4.24).

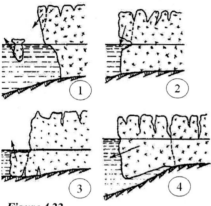

Figure 4.22
1) The sea water melts the glacier under the surface of the water, parts will fall off. 2) Waves can erode the glacier, smaller parts can fall off. 3) Ascending forces can press parts of the glacier against the surface. 4) The weight of the iceberg could break off large parts (iceberg) when the glacier no longer has support at the bottom.

When it starts to freeze in the autumn, new ice is formed. In sheltered areas this will be lying as fast ice. In its turn, this can break up as a result of environmental effects and tidewater differences. During this process, usually characteristic "finger-rafting" occurs (Figure 4.25) which can be the first stage in a zone where pressure ridges can occur. In the sea, the new ice will break up and drift about as drift ice (Figure 4.26). Some of the drift ice will pack in the Arctic Ocean without melting during the next summer.

Thereby there will be partially packed polar ice of multi-year ice. The partially packed polar ice will normally not be thicker than 3 - 4 metres, while in the pressure ridges it can often reach 10 - 15 metres (Figure 4.27 and 4.28). The pressure ridges will first be high and irregular (also known as "hummocks") and later "ground down" by wind and weather. The partially packed polar ice will mainly consist of multi-year ice, but will also have a considerable amount of one - and two-year ice.

Figure 4.23
Brash and slush is first a visible track of ice on the surface. This can create significant resistance for smaller ships with small engines. The cooling water intakes can also be very easily blocked.

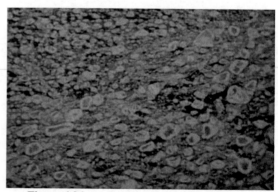

Figure 4.24
Pancake ice is formed as a result of newly-frozen floes packing towards one another.

Figure 4.25
Finger rafting in relatively thin and newly-frozen ice in Gulf of Bothnia.

Figure 4.26
Relatively open drift ice in the Kara Sea (4 - 5/10).

Figure 4.27
Old snow covered packed polar ice where the remains of pressure ridges / hummocks can be seen.

Dependent on current and wind, the drift ice can be in constant movement and how much of the surface it covers will usually be stated in tenths or per cent. If there is close drift ice we usually say that it covers 8-10/10. Under such conditions extremely large floes can be found of up to several kilometers in length. Close drift ice can be opened by wind and

current and channels are formed. In larger channels which will freeze again, thin even ice is formed – often called "polynia". By studying the fracture lines of large floes, experienced observers can see characteristic "melt layers" which can almost be regarded as year-rings. It is not unusual to find ice which is 5-6 years old in the Arctic Ocean (Figure 4.29). Gradually, as the climate becomes warmer the share of first-year ice is larger. In 2007, 59% of the ice was first-year ice. In 2008 the share had increased to 72%.

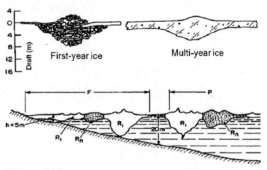

Figure 4.28
Different types of ice and pressure ridges. P = Partially packed polar ice, F = Solid/ fast ice, R_1= first-year pressure ridges, R_n= packed multi-year pressure ridges.

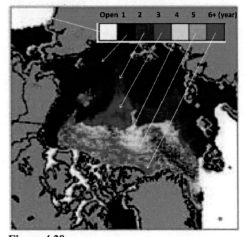

Figure 4.29
The colour codes show the age of the ice (March 2007). The oldest ice is found on the north side of Greenland and Canada (Source: NSIDC).

79

Glacial ice has about the same self-weight as sea ice. This ice is formed by snow which under pressure has become ice. The air in the snow is encapsulated in bladders under pressure. During melting these are released and a loud cracking sound can be heard. Pieces of glacial ice which remain lying in the water are therefore called "growlers" in (Figure 4.30).

Figure 4.30
Photo of growlers in front of a glacier at Novaya Zemlya. In open sea they can be difficult to see, and can cause extensive damage to ships.

Icebergs

When ice loosens from the glaciers, growlers and "bergy-bits" will drift away from the glacier front (Figure 4.22). When the bits are over a certain size, it is usual to call them icebergs. Icebergs can have various shapes and sizes, according to what the terrain and environment is like where the glacier falls into the sea. The largest icebergs are formed in the Antarctic, but from the large glaciers in the Arctic, one can also have icebergs of enormous dimensions. Dependent on the shape of the icebergs it is

usual to categorise them with different names (Figure 4.31). An iceberg will gradually melt and buoyancy and stability will be changed. This means that one can often experience icebergs that capsize, and possibly break up. When this takes place, it is extremely dangerous even for large ships to be in the close vicinity. Often one can also experience that the ice under the water reaches further out than what is visible. This is called an "ice foot", and can also be particularly dangerous for ships that sail close into an iceberg. During the period 1988 – 1992 a thorough study of icebergs in the Barents Sea was carried out, where transponders were installed on 56 different icebergs. Of the 1070 icebergs which were studied, it was calculated that the average weight was 300 000 tons, and that they drifted at a speed of 0.5 knots. The largest iceberg which was observed in this study weighed 6.2 million tons. It was also observed that the speed of drift could reach 2 knots. This drift normally follows the tidewater cycle and will move in elliptical paths (Figure 4.32).

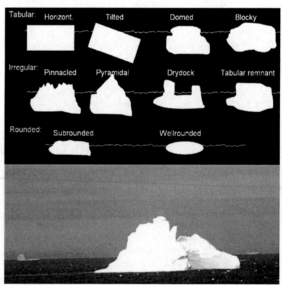

Figure 4.31
Different shapes of iceberg. Under a pyramid-shaped iceberg off the coast of Labrador.

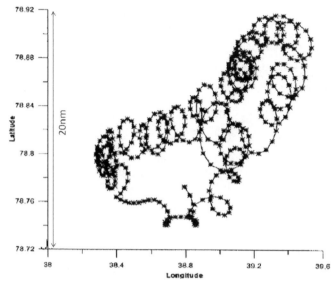

Figure 4.32
An example of a 3-day drift of an iceberg
in the Barents Sea. The plotting interval of
one hour is indicated by a cross (Source:
Norwegian Polar Institute).

4.5 The international "Egg Code"
and interpretation of ice charts

On the Norwegian and Swedish ice
charts which are described in
Chapter 2, ice information is
connected to the different patterns or
colour on the chart. This pattern is
normally self-explanatory in that the
symbolism is printed on each chart.
An example of this is shown in
Figure 4.33. Information on these
charts will then be connected to the
concentration of ice on the surface of
the sea. This can be given in % or
tenths. Exemplified:

- Open drift ice – 2/10 or 20%
- Close drift ice – 7/10 or 70%

In some cases the patterns can be replaced
by colours. The colour coding is an
international standard and will normally be
shown on the chart. On Russian charts,

different symbols are also used for
the different types of ice and
spreading of them (Figure 4.34),
but here also it is usual to use
colour codes.

If one is to obtain more detailed
information about the ice, such as
thickness, age, size of floe, etc.,
far more extensive coding is
required. In such cases it is usual
to use the "Egg Code". This is
usual in Canada, Denmark
(Greenland) the US and Russia.
The ice charts will then have a
reference to an ellipse ("Egg")
with different divisions and
numerals (Figure 4.35). The
ellipse is divided into four
different areas (lines) which describe
respectively:

- Total ice concentration
- Partial ice concentration
- Level of development (thickness)
- Size of the floes

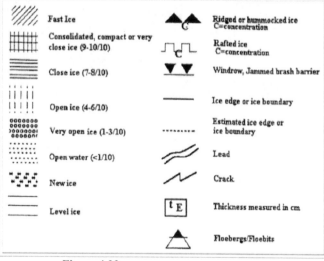

Figure 4.33
The symbols for different ice concentration
on ordinary Norwegian and Swedish ice
charts.

⌒	Giant floe, > 10 km	▲▲▲▲	Ridged ice
⬡	Vast floe, 2-10 km	◆◆◆◆	Ridged ice zone
〰	Big floe, 0.5-2 km	▲	Grounded ice, stamukha
◇	Medium floe, 100-500 m	▲▲▲	Grounded ridges in fast ice
○	Small floe, 20-100 m	•2	Snow cover ratio, max 3
▽	Ice cake, 2-20 m	⊛	Frazil, <5 cm
✳	Brash ice, < 2m	◇⊖⬡	Grey ice, 10-15 cm
1000	Fracture, width	◇	Grey-white ice, 15-20 cm
↯	Small crack		
→1←	Compacting ice field	◇ ///	Thin first-year ice, 30-70 cm
⎍⎍	Rafting		
⚠	Ridging ratio, max 5	◈ ///	Medium first-year ice, 70-120 cm
⦶	Compressed ice edge	◊ ////	Thick first-year ice, > 120 cm
〜	Diffused ice edge	△ ///	Second year ice
///	Fast ice	◆ ///	Multiyear ice

Figure 4.34
Symbols for ice concentration, type and spreading they are found on Russian ice charts.

If there is little information about a place, the "egg" will often be seen with only one numeral. This will then be the concentration in tenths. A detailed description of the symbols is given in Figure 4.36.

An example:
From the chart in Figure 4.35 we select the egg which is placed at the southernmost tip of Greenland. The explanation will then be as described in Figure 4.36:

1-3 means that there is open drift ice on the surface of the sea, and the concentration is between 1 and 3 tenths.

7.6 means that there is Polar ice (multi-year ice) with some thick winter ice (one-year ice).

3 2 1 ~9 means that the ice consists of small floes of less than 100m in length and that there are dense belt formations of ice floes where the concentration is 9 tenths.

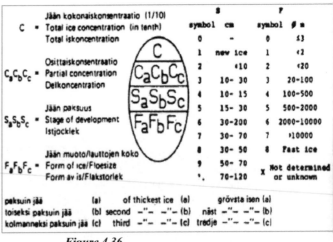

Figure 4.35
Part of an ice chart for Greenland where the Egg Code is used to described the ice conditions.

Figure 4.36
Detailed explanation of the codes which are used in the international Egg Code.

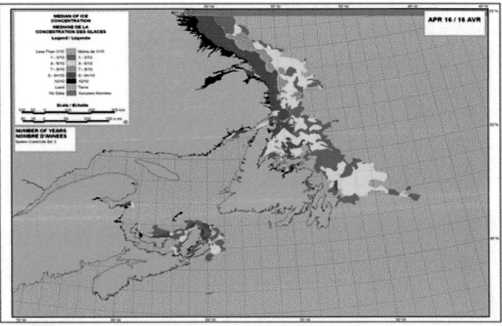

Figure 4.37
An example of a long-range ice forecast (Seasonal Outlook) from the area at Newfoundland as it is expected in mid-April 2010, but the forecast was late in the autumn of 2009.

In recent years several services have been set up for long-range predictions for ice coverage. This is based on meteorological and oceanographic conditions in the previous summer and autumn, as well as experience from previous years. Institutes in Canada and Russia have worked extensively with this, and in Canada a service has already been established on the Internet by the National Ice Centre (http://ice-glaces.ec.gc.ca/). Such long-range forecasts that can be issued several months in advance are called "Seasonal Outlook" and are presented in report form and with colour-coded ice charts (Figure 4.37). For those who are planning operations, these long-range forecasts can be a useful aid..

4.6 Questions from Chapter 4

1)
What does an ice molecule consist of?

2)
At what temperature will fresh water and seawater have the greatest density?

3)
Why will multi-year ice be harder than one-year ice?

4)
Why will snow on the ice influence the freezing process?

5)
How is the tensile strength in relation to compressive strength on ice?

6)
Describe what characterises the pressure from an ice floe which is pressed against a construction.

7)
What is meant by a "growler"?

8)
How do pressure ridges occur and how thick can these be in the Arctic Ocean?

9)
In which areas of the Arctic Ocean will we normally have the oldest ice?

10)
How large icebergs can we expect to find in the Barents Sea?

11)
On an ice chart of Canada you will find codes based on the international Egg Code. What does it mean when there is only the figure 9 in such symbol?

5 Technology for arctic shipping

The progress in the work with researching arctic regions had a close connection with the development of stronger ships – first in wood, later in steel. Most polar enthusiasts know the story about Fridtjof Nansen who gave Colin Archer the task of building en expeditionary vessel which could tolerate being frozen in the polar pack ice. The concept was that the ship was to be reinforced, as well as have a design so that the ice would push the vessel up. In addition it was to be possible to dismantle the rudder and propeller from the surface. The ship was named *Fram*, and was ready for Nansen's expedition in 1893 (Figure 5.1). During the three-year long expedition it was proved that Fram's hull form made it possible for a ship to survive extreme loads of heavy ice which was in constant movement.

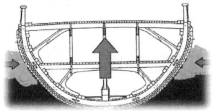

Figure 5.1
The Polar ship Fram was built with convex ship sides in order to be liftet out of the ice pressure against the sides of the ship.

Later, Det Norske Veritas (DNV) introduced new requirements for ships which could be exposed to loads from ice. In practice, this was a reinforcement of 15 – 25% of frames, plates and stiffeners, in relation to conventional class requirements.

5.1 Classification of ice-going vessels

The various classification companies lay down their requirements based on the conditions in which the ship is intended to operation, for example:
- Ice conditions and type.
- Temperature.
- Possibility of assistance from icebreakers.
- Special regional requirements.

Further, one can categorise the ice classes based on how contact with ice can take place:
- Constant movement in even ice.
- Contact with broken floes (i.e. in channels).
- Collision with ice edge /pressure ridge (more or less ramming).
- Pressure on the hull from pack ice.

The first two situations, which represent a relatively moderate load, form the basis for construction and classification of ice-going merchant and fishing vessels. This is often characterised as "Baltic" classes (in Canada, "type-ships"), and must normally be assisted by icebreakers in close drift ice. Typical vessels in this category are merchant ships which frequent Baltic waters in winter. Fishing vessels operating in waters at Greenland and Svalbard are in the same category. Fishing vessels that operate in ice can often have a differentiated reinforcement, i.e. the hull can be reinforced to class1A, while the propeller system can have class 1C (Figure 5.7). The reason for this is that propeller systems with a high ice class do not have the same favourable degree of fuel consumption and acoustic propensities as weaker and more optimised propellers. There will also be ships in this category that will be assisted through the

Northern Sea Route. In Det Norske Veritas (DNV) the Baltic classes will be designated with Ice 1A*, 1A, 1B and 1C (Table 5.1).

For ships who are to frequent the Northern Sea Route there is a requirement for an ice class minimum Ice 1A. The following areas will mainly be included in special requirements based on which class applies:

- Global and local strength of hull (plates, frame, web and other structural reinforcements)
- Machinery and cooling system
- Propeller and rudder system
- Heating of ballast tanks
- Mooring arrangements, etc.

Table 5.1
Overview of the ice classes and appurtenant areas of use in DNV.

DNV-notation	Equivalent Baltic class	Vessel type	Ice condition	Limitation
Ice-C			Very light ice	
Ice-1C Ice-1B	1C 1B	All type of ships	First year ice and broken channels 0.4m ice thickness 0.6m ice thickness	No ramming
Ice-1A Ice-1A* Ice-1A*F	1A 1A Super		0.8m ice thickness 1.0m ice thickness 1.0m ice thickness	
Ice-05 Ice-10 Ice-15		Icebreaking vessels	First year ice with ridges	
Polar-10 Polar-20 Polar-30		Main operation not icebreakin	Multi year ice With glacial inclusion	Accidental ramming
Icebreaker		Icebreaking is main purpose		Repeated ramming

The two latter ice situations will impose far greater loads on the hull. In order to operate under such conditions, the ships will have to be classified as **icebreakers** with an additional notation of "Icebreaker". In DNV these ships will be designated Ice-05,-10,- 15, Polar-10,- 20 and – 30 (Table 5.1). These ships will normally have special requirements beyond those which for the Baltic classes. These are mainly:

- Steel quality (low temperature)
- Extra bow and longitudinal reinforcement in order to be able to tolerate ramming and beaching.

The figures which are indicated behind ICE and POLAR indicate the ice thickness the ship can sail in with an even speed of 3 kts. With ICE-10 one could sail at 3 kts. through even ice of 1.0m thickness. For POLAR-20 the ice thickness can be 2.0m, etc. In this calculation of the icebreaker performance different ice strengths ice are used:

ICE-05	4.2 (N/mm^2)
ICE-10	5.6 (N/mm^2)
ICE-15	7.0 (N/mm^2)
POLAR-10	7.0 (N/mm^2)
POLAR-20	8.5 (N/mm^2)
POLAR-30	10.0 (N/mm^2)

Most classification companies, such as DNV, Russian Register of Shipping, Lloyds Register of shipping, Canadian ASPPR rules and the Finnish / Swedish rules, etc. will all have their own designations of their ice classes. At Table 5.2 is shown an overview of the ice classes in DNV, with corresponding equivalents from other companies. The comparison in the Table will not be equivalent for all the details, but in 2006 the international organisation for classification companies (IACS) agreed upon harmonised rules in this area (Table 5.2). This means that for all practical considerations one can act in accordance with given equivalents in the various companies and regimes. In addition to the rules connected to strength calculation which is described here, many classification companies have additional requirements connected to the temperatures in which the ship shall operate, but this will be further dealt with in Chapter 6. The harmonised rules issued by IACS are in regard to ships intended for polar operations, and are called *Polar Class* (PC). The operational criteria which are laid down for these ships are:

PC1: Operations year-round in all polar waters.

PC2: Operations year-round in moderate conditions with multi-year ice.

PC3: Operations year-round in second-year ice possibly with some multi-year ice.

PC4: Operations year-round in thick first-year ice possibly with some multi-year ice.

PC5: Operations year-round in medium first-year ice possibly with some multi-year ice.

PC6: Operations in summer and autumn in medium first-year ice possibly with some multi-year ice.

PC7: Operations summer and autumn in thin first-year ice possibly with some multi-year ice.

In DNV and many other classification societies, the development of modern class requirements are based on meeting the requirements stated in the Finnish / Swedish ice class rules from 1987. The division of classes will therefore have many similarities. The Baltic rules base construction load imposed by ice as an empirical factor between pressure and displacement / axis effect, where pressure, P, is assumed as:

$$P = f \bullet \sqrt{\Delta \bullet P_s}$$

Δ = max. displacement of the ship
P_s = max. horsepower on the shaft.
f = factor

Table 5.2
Table of equivalents for ice classes in different classification societies. In some sources the translation UL of the Russian letters YΛ will be found.

Baltic-classes / Type-ships	Class	High ice-class	<<<<	Class – notation	>>>>	Low ice-class
Det Norske Veritas	1A1	1A* / 1A*F	1A	1B	1C	C
Finnish / Sweedish rules		1A super	1A	1B	1C	
ABS	A1(E)	1AA	1A	1B	1C	II
Bureau Veritas	I 3/3 E	1A super	1A	1B	1C	
Germanicher Lloyd	100 A4	E4	E3	E2	E1	
Lloyds Register	100 A1	1AS	1A	1B	1C	1D
Polski Rejestr Statcow	KM	L1A / UL	L1	L2	L3	L4
Nippon Kaiji Kyokai	NS	1A super	1A	1B	1C	
Reg. of China	ZCA	B1*	B1	B2	B3	
Russian Register 1995	KM	ULA / UL	L1	L2	L3	L4
Russian Register 2003	KM	LU5 / LU4				
Registro Italiano Navale	100A-1.1	RG1*	RG1	RG2	RG3	
Canadian ASPPR		A	B	C	D	E
IACS PC		PC6 / PC7				
Icebreakers:						
Russian Register 1995	KM	LL1	LL2	LL3	LL4	
Russian Register 2003	KM	LU9	LU9	LU8	LU7	LU6
Det Norske Veritas	1A1	Polar-30	Polar-20	Polar-10 Ice-15	Ice-10	Ice-05
Lloyds Register	100 A1	AC3	AC2	AC1.5	AC1	
IACS PC		PC1	PC1	PC2	PC3	PC4/PC5

By static analysis of the damage to a range of ships, an estimate has been found for the response the plating and frame can resist. In calculating the plate thickness and frame dimension the strength calculations are based on the plastic bending moment and "construction ice pressure". From such empirical analyses there is also found plastic section modulus (W). Table 5.3 gives an impression of the ice loads in question, and how they are differentiated in class and ship's section. Figure 5.2 shows how DNV divides the ship up into different load sections. The principle of division will be similar in the various classification companies.

It will now be interesting to make a comparison of the various countries' rules. In Figure 5.3, for example, there is a graphical presentation of the classes' dimensioning of the forebody. The K factor referred to in the figures is based on the ship's displacement (Δ) and engine power (N):

$$k = \frac{\sqrt{\Delta \bullet N}}{1000}$$

Table 5.3
Calculated ice loads for the Russian ice classes as a function of the ship's displacement. Icebreakers are not included.

$\Delta(t)$	Region	Item			p (MPa)		
			Yλ A	Yλ	λ1	λ2	λ3
5000	Forward	Plating	2.92	1.65	1.23	1.00	0.84
		Frames	2.92	1.65	1.23	1.00	0.84
	Midship	Plating	1.82	1.03	0.65	0.43	0.27
		Frames	1.20	0.65	0.60	0.43	0.27
	Aft	Plating	2.19	1.03	0.52	0.27	0.14
		Frames	1.68	0.65	0.48	0.29	0.18
15000	Forward	Plating	4.72	2.67	1.99	1.62	1.36
		Frames	4.72	2.67	1.99	1.62	1.36
	Midship	Plating	2.36	1.33	0.89	0.57	0.34
		Frames	1.20	0.65	0.60	0.45	0.34
	Aft	Plating	3.30	1.33	0.72	0.36	0.17
		Frames	1.68	0.65	0.48	0.29	0.18
45000	Forward	Plating	6.01	3.40	2.53	2.06	1.73
		Frames	6.01	3.40	2.53	2.06	1.73
	Midship	Plating	3.00	1.70	1.14	0.72	0.43
		Frames	1.20	0.65	0.60	0.45	0.35
	Aft	Plating	4.21	1.70	0.91	0.46	0.22
		Frames	1.68	0.65	0.48	0.29	0.18

We see from the comparison in Figure 5.3 that the Baltic rules are low in relation to the Russian and Canadian classes, which are constructed to operate in arctic conditions. DNV's information indicates that their classes have a tendency to have thinner plating, but stronger frames in relation to the equivalent other classes. The reference in Figure 5.3 to "new Baltic" is equivalent to what was previously described as Finnish /Swedish rules of 1987.

Figure 5.4 shows the plate thickness of the Swedish icebreaker *Oden*, which is classed as "Polar-20" by DNV. Further, in Figure 5.5 is shown plate thicknesses for different parts of the hull on ships classed according to the Finnish / Swedish rules.

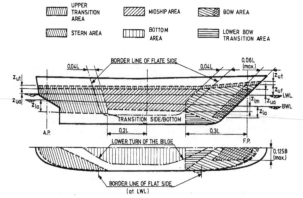

Figure 5.2
Divisions of ice reinforcement in the class regulation issued by DNV.

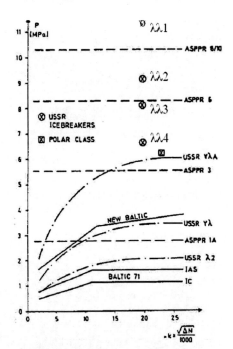

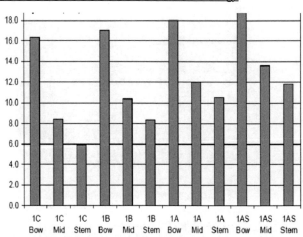

Figure 5.5
Minimum plate thicknesses (mm) in different parts of the hull of ships that are built to meet Finnish / Swedish (Baltic) rules.

Figure 5.3
A comparison of construction ice load on the forebody for different ice classes.

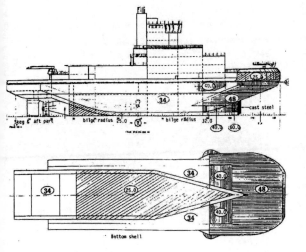

Figure 5.4
The figures indicate the plate thickness in mm. for the icebreaker "Oden", which is classed as "Polar-20" by DNV.

Examples of ships with different ice classes

In the following Figures 5.6 – 5.11 are shown examples of different types of ships that are built to ice classes in DNV. Pay special attention to Figure 5.6 of the trawler *Remøy Viking*. Here one has chosen to build a strong hull with a high ice class, but chosen a propeller and rudder system in a somewhat lower class. The reason for this is that one wishes to avoid the disadvantages in relation to lower efficiency, towing power and noise which a more powerful propeller will entail. When such a solution is chosen, it will naturally mean that one accepts the risk of having damage when one presses the ship to the design limit of the hull. This type of choice is something which is often made for trawlers that often operate in ice-covered waters.

Figure 5.6
The shrimp trawler Remøy Viking has the class notation +1A1, ICE-1C, (Ice 1A* for hull and Ice-1B for rudder), Stern Trawler, E0.

Figure 5.9
The express Coastal Steamer Nord Norge has the class notation +1A1 ICE-1C, Car Ferry A RM(-28°C/+30°C Sea) E0 F-C, NAUT-CPWDK.

Figure 5.7
The chemical tanker Trans Arctic has the class notation +1A1, E0, W1-OC, Ta for Chem and oil, Ship IMO type 2, Ice-1A.

Figure 5.10
The coastguard ship Svalbard has the class notation +1A1, Icebreaker Polar-10, RPSF-A, E0, Heldk-SH, Deice, Firefighter1.

Figure 5.8
The passenger ferry Color Fantasy has the class notation +1A1 ICE-1B, Car Ferry A MCDKCOMF-V(1) RP E0 F-M NAUT-OC CLEANPWDK TMON.

Figure 5.11
The research and offshore ship Polarbjørn has the class notation +1A1 E0 Icebreaker, Ice-05, HeldkICS, Dynpos AUT-R, Polar cl 10W1-dat (-30C), dk(+), pwdk ram.

Disadvantages of ice-classed ships

Building a merchant ship or fishing vessel with a high ice class has some clear disadvantages compared to conventional ships. Examples of these are:

- Higher building costs because of requirements for more steel and engine power. An increase of 20 – 30% must be expected. If we talk about a reinforcement of up to PC3 and a Double Acting concept, we are perhaps talking about an increase of over 50%.
- Lower loading capacity (deadweight) and freeboard because of heavier hulls. DNV has shown calculations that a typical 4000TEU large container ship will be 4, 8 and 12% respectively given reinforcement equivalent to PC5, 4 or 3. This will depend a little on the ship type and size.
- The hull will often be considerably less optimised for speed in open waters. This entails higher fuel costs. The same problem will be connected to a heavily reinforced propeller.
- Heavy hull leads to a great metacentric height, which in its turn will give the ship intense and rapid rolling movements in open waters. The fact that bilge keels are excluded on ice-going ships contributes to increasing the problem. Special hull forms and propulsion systems could also contribute to more pitching and noise in open waters.
- Ice class is valid only with defined draught.

5.2 Icebreakers

Icebreakers are the backbone of shipping along the Northern Sea Route and in Baltic water. In addition to safeguarding shipping, the icebreakers in the northern areas are important in relation to sovereignty and emergency preparedness.

It is first and foremost Russia and Finland which in their model tanks have developed ships for the Arctic. Canada, the US, Japan, Germany, Sweden and Norway have also come far in this development. In particular, there has been a lot of research on hull design and fuel economy. Many of the states mentioned have freezing capabilities in their model tanks (Figure 5.12). In the 1970s in Norway an attempt was made to simulate ice using the wax covering on the surface of the model tank in Trondheim. The attempt was not particularly successful and has not tempted repetition.

Figure 5.12
Model tanks are important for testing of ships and constructions which are to operate in ice. Here, HSVA in Hamburg.

Important requirements which should be made to an arctic icebreaker are as follows:
1. High icebreaking capacity.
2. Good manoeuvring characteristics in ice.
3. Good towing characteristics.
4. Little ice in the channel behind the ship.
5. Long operating radius.
6. Environmentally sound.

Items 1 - 4 are further described later on. Item 5 is extremely important from a safety point of view. If ice conditions should occur which make sailing difficult, it is of decisive importance that the icebreaker can operate continuously without having to interrupt operations to bunker. It can be a long distance to the nearest bunkering station in the Arctic. Ordinary icebreakers, however, do have a dilemma regarding this point. If at the same time there are to be powerful engines and a long operating radius / time, the ship will necessarily be regarded as an icebreaking "tanker". In order to solve this dilemma, the Russians have concentrated efforts on nuclear powered ships which in theory are built to be independent of fuel supplies for <u>four years</u>. It has therefore not been necessary to develop large bunkering facilities in Siberia.

The performance of the nuclear icebreakers was also demonstrated in the Chuchi Sea (Siberia) in1983 when 51 merchant ships were trapped in unexpectedly difficult ice conditions. Had it not been for the three nuclear icebreakers the Soviet Union had at that time, this situation could have been one of the worst shipping catastrophes to date. Only one ship was lost in the operation. Another episode took place in 1993. At that time, the Russian diesel icebreaker *Kapitan Dranitsyn,* which was on a passenger cruise around Greenland, was trapped in the ice in the Canadian Arctic between Greenland and Ellesmere Island. The situation was very critical and none of the local icebreakers could come to her aid. The nuclear icebreaker *Yamal* therefore sailed over the Arctic Ocean and assisted the casualty back to more open waters (Figure 5.13).

Item 6, being environmentally sound, is a very important point since the environment in the Arctic is very sensitive and is the growth area for enormous fish resources. The problem will be further discussed in Chapter 7, but it must be emphasised that in practice, the best possible compromise must be sought between items 1-5 and item 6.

5.2.1 Hull design, load and ice resistance

The design of the bow of an icebreaker has two main goals. The most important one is that it is able to break the ice in an efficient way. This is traditionally done with a very extended bow which means that the ship slides up on the ice and breaks it down. Further, the bow must remove the broken ice out to the sides, to avoid that it comes under the bottom. Ice floes which come under the bottom will represent a great danger to propellers and rudder. In addition, it will be an advantage for following ships that the channel is as open as possible. An example of such "two-stage" bow is shown in Figure 5.14.

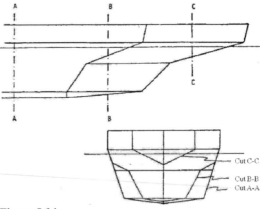

Figure 5.14
***Idealised bow for ice-going vessels. The ice
is broken in cut C and pushed aside in cut
B (Herfjord, 1982).***

Figure 5.13
***The nuclear icebreaker Yamal escorts the
diesel icebreaker Kapitan Dranitsyn
through difficult ice.***

In order to gain a little better understanding of the loads on the hull it is necessary to go into the theory. The reference system is defined as shown in Figure 5.15. If we imagine that the ship sails into a thick ice floe, it will be appear as shown in Figure 5.16. Two types of loads will occur on the ship. First and foremost locally in the bow region, but also globally in the form of a bending moment and shearing forces. The local loads can be found by multiplying the area of the imprint by the breaking strength. The breaking strength of the ice can vary according to the type of ice which is involved. Newly frozen ice is relatively soft, while old Arctic ice is much harder (blue ice).

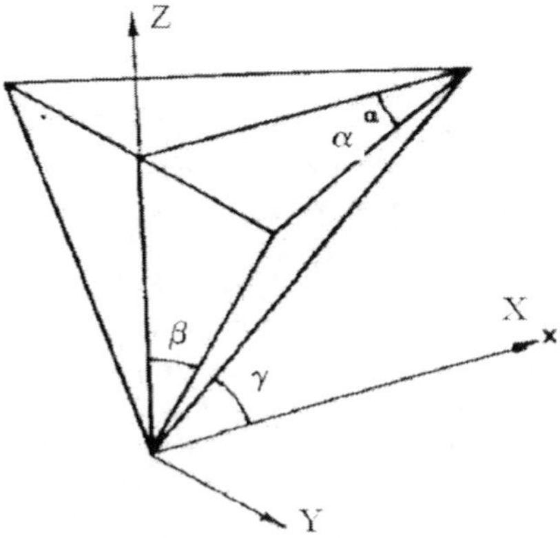

If we assume that the ship's vertical forces were initially evenly distributed along the whole of the ship's length, after the collision we will have vertical forces and moment as shown in Figure 5.17. If we do a numerical analysis of the forces in the collision, we will obtain a course of events that is shown in Figure 5.17. Ships that sail at a constant speed through thick ice will also be exposed to large variations in loads. An example of this is shown in Figure 5.18. The figure shows the percentage (of max.) load on frame and stiffeners during sailing in relatively thin ice.

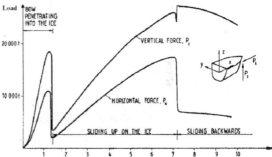

Figure 5.17
Load on the bow during collision with ice (ramming) as a function of time (Herfjord, 1982).

Figure 5.15
Reference system for idealised bow.

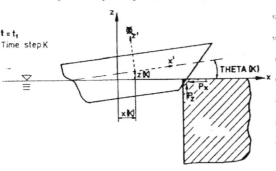

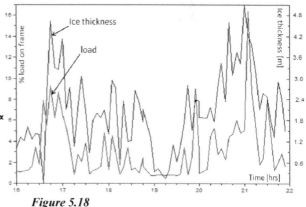

Figure 5.18
Load on the hull when KV Svalbard sails through different ice (Larsen, 2008).

Figure 5.16
Movement of ships that enter a larger ice floe.

The following data is used in the calculation basis for Figure 5.19:

Length, L= 240m,   Width ,B = 40m
Displacement   = 90,000t
Speed (ahead)   = 15 kts
Breaking strength (ice)   = 700 t/m^2
Dynamic friction coeff.   = 0.15
Static friction coeff.   = 0.35

The breaking strength of the ice which is chosen in the example is typical for 2m thick ice which is not older than 2 years. It is, of course, not insignificant how the bow is designed in order to calculate the max. load on the bow. In Figure 5.19 is shown the horizontal and vertical max. forces as a function of the design of the bow. When the gliding movement forward is stopped, the ship will either be stuck fast, or slide back. If the ship is stuck fast and cannot free itself with its own engine power, trimming of the ship will be necessary. Icebreakers have enormous pump and trim / heeling systems which can be used in such situations.

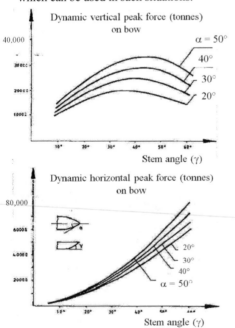

Figure 5.19
Maximum load forces as a function of the design of the bow.

In Figure 5.20 is shown the forces necessary to pull the ship loose. The alternative is to trim the ship astern to slide off. Ships with azimuth propellers will also be able to flush large volumes of water along the sides of the ship. This will result in that the ice is weighed down and the ship will be freed more easily.

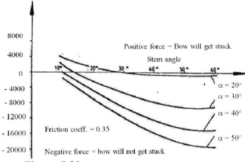

Figure 5.20
Traction force or trim weight which is necessary to free a ship (ship data above) that is stuck fast on an ice floe.

The icebreaking capacity depends on the strength of the vertical pulsation .The powerful pulsation in the collision, which lasts 1-2 seconds, however, is not enough. This is because the deformational response of multi-year ice is 10 - 20 seconds. Consequently, icebreakers must seek a large vertical force throughout the whole of the gliding process. Figures 5.19 and 5.20 show that the capacity will decrease with increasing bow (stem) angle. The horizontal bow angle is of lesser significance.

Figure 5.21 shows the modification of the bow of the Russian icebreaker "Kapitan Nikolajev", where the bow angle was reduced to increase the icebreaking capability. After modification, considerably better icebreaking capacity was shown, and a considerably more open channel, which is one of the most important qualities of an icebreaker. It should also be mentioned here that not only was modification a reduction of the bow angle, but also a change in the design itself, to a more spoon-shaped curvature.

Figure 5.21
The Russian "Kapitan Nikolajev" with a sketch of the new bow (shaded). The ship was modified in 1990.

The disadvantage of pointed bow angles, however, can be that one has a ship which has bad characteristics in open sea. This is a circumstance which is especially important for icebreaking offshore vessels which will normally operate in open waters for more than six months of the year (Fig.5.22). The Norwegian coast guard ship Svalbard is another example of a hull form where the fact that the ship must be able to sail at high speeds in open sea has been taken into consideration (Fig.5.23).

As will be realised, there will always have to be a compromise between the hull load and the icebreaking capacity. On Russian icebreakers it has been usual to use a bow angle of approximately 20 degrees and a horizontal angle of approximately 30 degrees.

Stern and towing arrangement
The design of the stern is also important for the icebreaker to be able operate optimally. It must first and foremost be designed to let go of the ice. This will make the channel more open for following ships, while at the same time the ice friction will be reduced. The risk of ice load on the propellers will also be reduced.

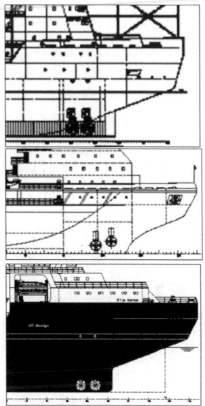

Figure 5.22
Bow design on different icebreaking offshore vessels. Top: Moss 828 Ice, Middle: Fesco Sakhalin, Bottom: Rolls-Royce UT-758.

Figure 5.23
The bow of the KV Svalbard is relatively pointed to make the vessel better in open sea.

In assisting ships and in ramming, icebreakers will be dependent on being capable of backing efficiently. Such operations can be critical for rudder and propeller, but it is also essential that the ice resistance is small, so that the ship is capable of backing efficiently.

A third characteristic feature of the stern of icebreakers is the towing notch at the stern (see Figure 5.24). This is a heavily fendered area which is designed with the intention that a following ship shall be able to put the bow into it in towing operations. The area is exposed to large loads and must necessarily be designed accordingly. For the towed ship it can often mean deformation of plates in the bow if the attachment is not done with the greatest care. This problem occurs most often when the notch does not meet the ship's ice-reinforced area, as for example, with heavily loaded ships. By using the towing cleft it is also possible for two ships to move together in "tandem operations", where the forward ship pulls and the one astern pushes. Such operations are not unusual with the SA-15 ships operated by Russia along its Northern Sea Route.

Figure 5.24
Icebreakers in Finland, Sweden and Russia are equipped with towing notch in order to be able to perform "close-couple towing". Here of the icebreaker Nordica.

More and more ice-going vessels are now built with pulling azimuth propellers in order to move with the stern towards the ice ("double-acting"). This requires that the stern is also designed to break the ice effectively, which means that the problem which was discussed under bow design is also highly relevant to the stern (Fig.5.25).

Figure 5.25
The stern of the Russian icebreaker Fesco Sakhalin with two azimuth thrusters (ABB Azipod) which are built for operations at Sakhalin (Source: Aker Arctic).

For ships with propellers and rudders a special arrangement is required to protect the rudder. For ships with one conventional rudder this will normally be a fin (ice-knife) that protrudes behind the rudder when this is in the centre position (Figure 5.26). In such cases it is important that the rudder is set in the centre position when the ship moves astern in the ice. If the ship has two rudders, there can be an arrangement to set the rudders against one another like a plough when the ship reverses. In order to prevent loads on the rudder, a block has been made on which the rudder rests. The icebreakers *Oden* (Figure 5.27) and the three *Viking* ships (Thor-, Vidar- and Balder Viking) have this arrangement. Such "plough" will also protect the propeller against ice.

Figure 5.26
A fin (ice- knife) in the hull protects the rudder when reversing. Here on the bulk carrier Arctic.

Figure 5.27
The stern of the Swedish icebreaker Oden. The ship has two variable pitch propellers and the two rudders are placed automatically in the "plough" position when the ship reverses.

"Double Acting Ship" design
The conflicts which are described above concerning optimum design for ice or open waters has always been a great challenge for ship designers. In addition to the challenges by designing an optimum bow for icebreaking, one will always have to take into account the fuel economy of the ship for the whole year. Just think of how great the savings will be on a tanker to use a relatively large bulb in the bow.

Unfortunately, a bulb will not be suitable for icebreaking. Also, Russia would normally not accept bulbs for voyages along the Northern Sea Route. After the introduction of the azimuth thrusters, including for large tankers, a new idea surfaced – the "Double Acting Tanker" or "Bi-directional tanker". This is a design where the bow is a conventional bulb bow, and where the stern is designed for icebreaking (Figure 5.28). Thereby, the best of two worlds is achieved. Trials and experience from such tankers in difficult conditions in Northwest Russia have given extremely good results and contribute to other types of ships being given this design.

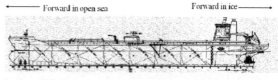

Figure 5.28
Double Acting Tanker is built to move forward in open waters and astern in ice (Source: Aker Arctic).

Ice resistance.
When the ship sails in relatively spread drift ice, less than 5/10, it is not usual to calculate substantial ice friction. When the concentration of drift ice increases, the resistance will increase rapidly. If the ice is moving or under pressure, the resistance can be greater than if it was solid, continuous ice of the same thickness. It is usual to regard the total resistance as the sum of:
- Breaking of the ice in the bow region.
- Pushing down the broken ice.
- Friction between the ice and side of the ship (Figure 5.29).

In the area between 5/10 and 10/10 cover of drift ice, we can make a rough estimate of the ship's speed by using the following formula:

$$V_p = V_0 + (V - V_0)*((m - 0.5)/0.5)$$

V_p = speed in partial cover (between 5/10 and 10/10)
V_0 = ship's speed in open waters.
V = ship's speed in even, solid ice.
m = densely packed ice cover over 5/10

It is assumed in the formula that the ice is not under pressure.

Example:
The Russian icebreaker "Mudyug". Speed in open waters is 16 knots. Speed in 0.9 m. even ice is 5 knots. Ice cover is 8/10.

$$V_p = 16 + (5 - 16)*((0.8 - 0.5)/0.5) = \underline{9.4 \text{ knots}}$$

If a more accurate calculation is to be done, it is a complex problem where hull form, ice type and not least whether there is snow on the ice will play a substantial role. A complete analysis will consequently give a formula with constants adapted to ship and conditions. As an example is taken an empirical formula which is based on measurements for the Russian icebreaker "*Kapitan Belousov*". The formula indicates total resistance (R) in metric tons:

$$R = 8.6 \bullet L^{0.5} \bullet B^{0.25} \bullet \left(0.4 + 8 \bullet \left(\frac{V}{\sqrt{g \bullet L}} \right)^{1.25} \right)$$

L = length of ship,
B = width of ship,
V = speed.
g = gravitational acceleration

The friction when the ice glides along the side will be very dependent on whether there is snow cover. Snow cover will make the resistance considerably greater.

In the construction of ice-going vessels and icebreakers it is of great interest to reduce

the size of both types of resistance. In the next chapter we shall discuss more in-depth how this goal can be achieved.

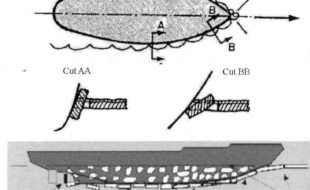

Cut AA   Cut BB

Interaction   Friction   Submerging   Breaking

Figure 5.29
Calculations of the ship's resistance is very dependent on how the ice follows the side of the ship.

5.2.2 Aids for reducing ice friction

The bow design will be especially important to reducing the resistance in icebreaking itself, as well as the energy which is used for pressing the ice down (submerging). In recent years, new vessel concepts have been developed to increase the icebreaking capacity and reduce fuel costs. This means that a range of different designs and aids have been tested. In this development there are mainly two types of forebodies which stand out from more conventional bow forms. This is the typical spoon form, which is also used on the Swedish icebreaker *Oden* (Figure 5.30) and the German Thyssen / Waas construction, where the bow reminds more of a double plough share (Figure 5.31 and 5.32). The spoon-formed and strongly forward-leaning bow was developed for Canadian supply ships in the 1980s and further developed in different variants both in Sweden and Finland, and where *Oden* is

possibly the most exaggerated example.
There is no doubt that the bow is effective in
ice (Figure5.36), but when the ship sails in
open waters there will be a lot of strikes
against the bow – with resulting loss of
speed and discomfort on board.

Figure 5.30
*Three different ships with different spoon-
shaped bow design. Above, the Canadian
Canmar Kigoriak (later Vladimir
Ignatyuk), in the centre the Swedish Oden
and at the bottom the Finnish Nordica.*

Common to all these ships is that they have
a form of plough at the bottom where the
bow transitions to the keel of the ship. The
function of this is to push heavy ice out to
the side so that the bottom plates are not
exposed to great loads. On some ships there
has been ice damage to bottom plates which
are not reinforced to the same degree as the
sides of the ship. On offshore vessels the

plough is designed in such a manner that
one can place tunnel thrusters there (Figure
5.22). Such thrusters will be dimensioned
for ice and are normally used only in open
waters.

The other concept which was mentioned
introductorily was the German design by the
Thyssen/Waas yard. Modification of the
Russian sub-arctic icebreaker *Madyug*
(1986) (Figure 5.31) and the arctic
icebreaker *Kapitan Sorokin* (1990) (Figure
5.32) by Thyssen/Waas showed that with
the new bow the fuel consumption during
icebreaking was reduced by up to 65% in
ideal conditions. This will naturally entail a
far greater operational radius for the ships. If
this is brought into the discussion regarding
the requirement to use nuclear icebreakers in
the Arctic, it can create interesting
perspectives. Nevertheless, there is a long
way to go before one can compete with
Russian nuclear vessels which operate for 3
– 4 years between each "bunkering" of
enriched uranium.

Figure 5.31
*The Russian Madyug during modifications
at Thyssen-Waas. Note the sharp "plough
share" on each side of the new bow.*

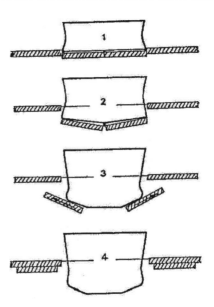

Figure 5.33
The principle for icebreaking according to the Thyssen/Waas concept. Note that the broken ice is pushed under the ice on each side of the ship.

Figure 5.32
The Russian Kapitan Sorokin after modification. The bow performs well in even ice, but creates certain problems in the varied arctic ice.

The main motivation for the new hull shapes was to increase the ice-breaking capacity considerably, at the same time as great emphasis was laid on being able to make the channel as open as possible (Figure 5.33). Regarding *Oden* it is calculated to be able to break 1.8m thick ice at 3 knots. This is an increase of 0.6m compared to the conventional hull shape which was previously built for operations in the Gulf of Bothnia (Figure 5.34). "Oden" is classified as "Polar 20" by DNV. The capacity of *Madyug*, which is good deal smaller than *Oden*, is also shown in Figure 5.34. In addition to having great importance in convoy operations, an open and marked channel will be better suited to loading and unloading operations. This is important on the Northern Sea Route since because of the depth conditions, trucks must often be driven out to the ships (Figure 5.35).

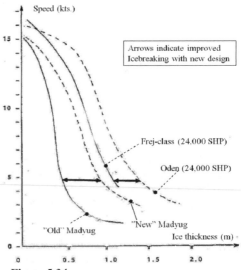

Figure 5.34
The "HV curve". The icebreaking capacity of Oden, compared with the older Frej class. The arrows indicate increase in the capacity. The curve also shows the capacity of Madyug, before and after modification.

Figure 5.35
Unloading on the ice from Russian icebreakers and merchant ships at the Yamal Peninsula in the Kara Sea.

However, in the modification of ships such as the *Kapitan Sorokin* according to the Thyssen/Waas principle, there has been an undesirable "side effect". The ship has been given slightly worse characteristics in rough head seas, as well as strongly varying arctic ice is not always tackled as well. The ship is not as manoeuvrable in the ice as it was before with the conventional bow. The ship operates therefore now (2011) mainly in Gulf of Finland where there are considerably easier conditions in winter. The *Madyug* is initially a somewhat smaller sub-arctic icebreaker which is now used in the White Sea and similar areas with one-year ice.

Other technical aids
Common to many of the new vessels is that a range of different technical and design-related aids are used to improve ability to navigate and the characteristics.

- **Water flushing**
As mentioned previously, the snow cover on the ice will result in considerably increased friction along the hull. Therefore, on many ships water pumps have been installed which flush the snow cover and thereby reduce the friction. The water comes out of a series of nozzles in the bow (Figure 5.36), but it is also important that the water jets reach so far out to the sides that they

effectuate reduced friction along the sides of the ship. The system is most effective when the ship moves in solid / even ice with a thick layer of very cold snow. When the ships are working in areas with a lot of channels and varying ice, the system is not used much. On some ships the nozzles can be operated in groups so that they can also be used as a bow thruster. An example of such control system is shown in Figure 5.37.

Figure 5.36
The Canadian icebreaker Robert Lemeur with flushing in the bow. Below is shown the nozzles on the Oden.

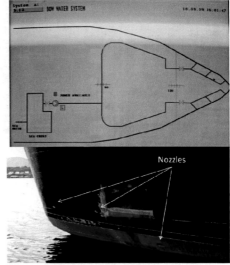

Figure 5.37
Control system for water jets in the bow of AHTS Vidar Viking.

- **Bubble system**

We know that in many marinas there is the system where air is released under the boats so that ice does not reach the hulls. This principle has also been taken into use on a range of ice-going vessels. The system was developed in Finland, and is based on large amounts of air being pumped out through nozzles which are placed under the waterline along the hull (Figure 5.38).

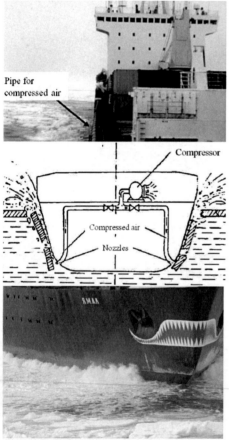

Figure 5.38
The bubble system to reduce ice friction along the sides of the ship. The photo shows branch pipes for air distribution on Russian SA-15. At the bottom we see the nuclear icebreaker Yamal with the system activated.

In this way, air and water circulates along the sides of the ship, which prevents the ice from reaching the hull and creating friction. If the ship is in open waters, it can also use the bubble system as side thrusters when manoeuvring– then the one side of the supply is shut off from the control panel on the bridge. The system has proved to be especially effective in solid ice and has been taken into use on several Russian and Finnish merchant ships and some icebreakers, including being installed on the nuclear icebreakers of the *Taymyr and Rossia* class.

- **High gloss polishing**

The hull of a ship which sails in heavy ice can in many ways be compared to a sled or ski that glides on ice and snow. It is therefore reasonably obvious that one must attempt to make the gliding surfaces (the hull) as smooth as possible. It is therefore usual that the subsea hull and the waterline (ice belt) are treated with a two-component epoxy coating (Inerta-160) since ordinary bottom coating will be worn off after a short time in the ice. Another variant is to use smooth stainless steel for the bow and ice belt. This is naturally relatively costly, but is used with good results on the newest Russian nuclear icebreakers. On nuclear icebreakers, this is heated by the cooling water in addition.

- **Heeling system / "Duck-walk"**

Even on the old sealers one used more and more sophisticated methods for heeling the ship. This was done to free the ship from the ice. On most of the modern icebreakers, large tanks are built in with pumps installed in-between to use the method when the ship sails in heavy ice or has got stuck. The principle is shown in Figure 5.39, and for *Oden* the tanks are 800m^3. An enormous pump, which in principle is a tunnel thruster with a variable pitch propeller, pumps the water from side to side in 15 seconds. This displacement of ballast water causes the ship to heel 7° to each side.

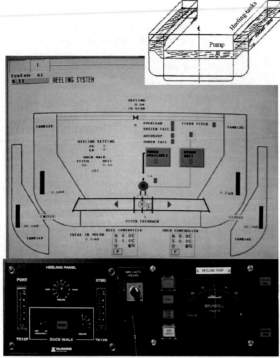

Figure 5.39
The heeling system on the icebreaker Vidar Viking. At the top is shown a principle sketch. At the bottom is shown the operator's panel on the bridge where the pump can be put into the"duck-walk" mode.

Both on *Oden* and the newer offshore icebreakers which are used in the Gulf of Bothnia the system can be put into an automatic and continuous mode which makes the ship "roll" forward in difficult ice. This is also called "Duck-Walk". On ships which have relatively large rudders, a similar effect can be achieved by quickly moving the rudders from side to side when moving forward in heavy ice.

As a curiosity can be mentioned that many years ago, the crew ran from side to side to try and free the ship from the ice (Nansen, 1924). Otheres have used a weight on the boom and swung it from side to side.

- **Reamers**

When the *Oden* was built, the hull was equipped with "reamers". This is a protruding construction which increases the ship's width just aft of the bow (Figure 5.38). The reamer prevents the ice attaching to the long part of the ship's side. This will entail reduced friction and better performance when the ship moves forward in the ice. If the ship heel s in a turn the reamer on the high side will to a certain extent come out of the water and the reamer on the low side will increase the width further. This in its turn will contribute to a more effective swing. Model tests on the *Oden* in 0.8 m thick ice show that by using the reamer at a 7° heeling angle, the turning circle radius is reduced from 2.5 to 1 ship's length. This characteristic means that the construction is often called a "turning-reamer". The principle is also taken into use on the combined icebreakers and offshore vessels that operate in Finland in the winter (Fennica, Nordica and Bothnica) (Figure 5.40) and in the Thyssen/Waas design which has been described earlier. A reamer can also have certain disadvantages. The extension represented by the reamer can create certain problems when mooring at a quay, as well as the positive characteristic is not present when the ship is backing in ice.

- **Propeller water control**

In easier ice conditions where there is only first-year ice, as in the Gulf of Bothnia and the St. Lawrence Bay, for example, propellers are also used in the bow. The propeller water from the propellers in the bow, in addition to pulling the ship, will contribute to reducing the ice friction along the sides considerably, because the propeller wash will push the ice floe away from the hull. Ice breakers with this propulsion method are also very manoeuvrable, but will also be vulnerable and little suited to arctic ice. Figure 5.41 shows the Swedish "Atle" class which is built with four propellers, two forward and two aft. At present (2011) there are three icebreakers of that class in Sweden and two in Finland.

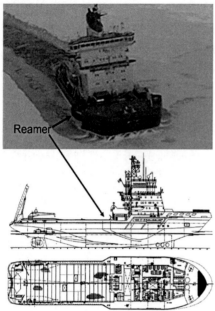

The advantageous effect of flushing propeller water along the ship's sides is also used on ships with azimuth thrusters which move aft in the ice. Icebreakers with this type of propulsion in actual fact move much better astern than ahead in some types of ice. This was also the case for the Norwegian KV *Svalbard* which is equipped with two pulling ABB Azipod propellers each of 5MW (Figure 5.42).

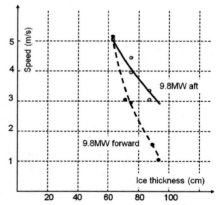

Figure 5.40
Top: The icebreaker Oden. Note the transverse bow with the "turning reamer" behind. Bottom: Reamer on the icebreaker and offshore vessel Nordica.

Figure 5.42
The figure (HV curve) shows performance ahead and astern of KV Svalbard during movement in even ice (Kilde: Coastguard / MARC)

Future ship types

It is the oil and gas activity in the arctic regions that has awakened the most interest for development of new ice-going vessel concepts. An early example of this was the American, ice-going tanker *Manhattan*. The problems with regard to large capital costs and the risk of environmental catastrophes have meant that the development of ice-going tankers has previously been of relatively little interest. With increasing oil prices and a petroleum activity which is steadily moving northwards, however, the picture is about to change. Therefore, tankers carrying a high ice class have already been built or contracted for, and many of them have a so-called "double-acting" design. The fact that all new tankers shall be built with a double hull has also

Figure 5.41
Finland and Sweden operate a series of icebreakers of this type. The ships are 104.6 m long and have 22 000 SHP. Propulsion is diesel – electric with 4 fixed propellers – 2 foreward and two aft.

contributed to that the price difference for tankers with and without ice classes is reduced. In the 1980s there were projects in the research environments which studied trans-polar transport of LNG and petroleum products. In that connection, Thyssen/Waas and others performed studies of LNG ships with the same bow design as described previously in this chapter. The focus was on relatively large ships which were designed to operate in tandem, as sketched in Figure 5.43. This was motivated by the Finnish-built SA-15 ships which were also designed for such operations. Since then, the development has continued, and LNG development has been adopted in several places in Russian areas, where one must expect ice in the winter. Therefore, there are contracts already in place for large ice-classified tankers for this traffic. For the ships which are intended to be operated in the most difficult areas and times of year, it is most probable that "double-acting" hulls with azimuth propulsion will be preferred (Figure 5.44).

2 x 75,000m³ LNG-tanker i tandem (beregnet 3m is med 7 knop)

140,000m³/75MW LNG-tanker

Figure 5.43
Thyssen/Waas LNG tanker concept, which was intended for trans-polar routes.

Figure 5.44
Probable design for transport of LNG from Russian gas fields (Source: Aker Arctic).

In order to make the tonnage as cost effective as possible, it could also be an idea to use smaller vessels which would perform a shuttle service for larger vessels and be lightering stations in open and protected waters. This is already routine to a great extent for shipping of oil from Northwest Russia (Figure 5.45), but technology remains to be developed for such operations, including for LNG.

Figure 5.45
Lightering of oil is usual in order to utilise the tonnage better. Here an example from Varanger in Finnmark (Source: Norwegian Coastal Administration).

Another market which is also growing is the cruise traffic to new and exotic areas. Expedition tourism is therefore both a possibility for utilising the icebreaker fleet in an otherwise quiet summer season, as well as a market for specially built smaller cruise ships. Possibly the most extreme form of such activity is operated by the Russians who each year send tourists to the North Pole with nuclear icebreakers (Figure 5.46). Smaller, comfortable vessels carrying an ice class are built for a larger market for voyages to the Antarctic, Greenland, Svalbard, etc. the Norwegian Coastal Steamers (Hurtigruten) has developed this type of traffic, and with its latest vessel *Fram* will further develop this market (Figure 1.9). Ships in this market will most probably have a relatively conventional design and moderate ice reinforcement in the future.

Figure 5.46
Since 1990 Russian nuclear icebreakers have transported tourists to the North Pole. Here from the grill party at 90°N.

Based on the great expectations regarding increased traffic off the northern coast of Russian, an extensive renewal programme has been drawn up for the icebreaker fleet. When the last of the ships which are operated today (2011) are expected to be obsolete about 2020, a series of three different new classes of ship are planned to be in operation. These are called LK-60, LK-25 and LK-18 (Perisypkin & Tsoy, 2006). The design is expected to be relatively conventional, but without doubt the experiences of the past 100 years of operation of polar icebreakers will be built in.

LK-60 will have nuclear propulsion of 60MW and have a length and width of 176 and 34m respectively. Displacement and icebreaking capacity is planned to be 32 400t and 2.9m respectively.

LK-25 will have diesel – electric propulsion of 25MW and have a length and width of 139 and 30m respectively. The displacement and icebreaking capacity is planned to be 19 500t and 2.0m respectively.

LK-18 will have diesel – electric propulsion of 18MW and have a length and width of 118 and 29m respectively. The displacement

and icebreaking capacity is planned to be 15 900t and 1.6m respectively.

Of a more spectacular nature asymmetrical hull and subsea vessels are also launched as the ideal transport means in ice-covered regions (Figure 5.47). An asymmetrical hull can be pulled forward abeam by azimuth propellers. A relatively small vessel will therefore be able to break a broad channel. Such ships can especially be used as tugs, icebreakers and tender vessels in ice-covered harbours. Such smaller vessels can also be used as contingency preparedness vessels for oil spill.

Figure 5.47
Asymmetrical hulls have been launched as a probable design for smaller icebreakers (Source: Aker Arctic).

Subsea vessels are more controversial, and the costs and environmental questions connected to such traffic will probably make projects impossible to realise. Even though the Arctic Ocean is very deep the depth conditions along the coast in the Arctic are not especially suitable for subsea vessels. Some people are also of the opinion that the Arctic Ocean is well suited to the use of ecranoplan (Wing In Ground). This is a form of aircraft which is lifted off by an air cushion and thereby does not fly higher than 10 – 20m. The Russian defence has built such vessels for transport of materiel (Figure 5.48). Even though this is also a design which is possible to realise, high costs and limitations will make the concept difficult to realise in the near future.

Figure 5.48
Russia has built an "ecranoplan" (WIG)
which many regard as the transport means
of the future in ice-covered regions.

5.2.3 Engines and propeller installations

Engines
Requirements for the engines themselves, whether they are diesel engines, turbines or nuclear reactors, are normally kept outside what is connected to the ice regulations of the classification companies. The only thing that requirements are made for in this manner is a minimum power output for the shaft, which is somewhat higher than the requirements for ordinary classes. In addition, there will be a number of requirements for support systems, such as cooling system etc.

Diesel-electric propulsion
Most of the arctic icebreakers are equipped with diesel-electric (D/E) propulsion. The

advantage of this is that one can have flexible control systems which will make the ship easily manoeuvrable, at the same time as the diesel engines are loaded in a favourable manner. There will also be more freedom in designing the engine room. However, there has been a general opinion that D/E propulsion is not economical because of the loss in the generators and electro motors. Several combinations of alternating and direct current generators and motors are used and have typically different degrees of effect:

Generator / Motor:	loss:
DC / DC	20 %
DC / AC	18 %
AC / AC	15 %

DC = direct current, AC = alternating current

On newer ships, both the generator and electromotor have AC, and in many cases will be equipped with frequency converters for better control of revolutions and reversing. How the electro motors are located will depend on what type of propeller system is chosen. On the really large icebreaker, such as the Russian icebreakers, the electro motors will be directly connected to the propeller shaft. On these ships there are normally three propellers without variable pitch, but where the blades can be changed in the case of damage (Figure 5.49). Steering control of these icebreakers is done by one rudder behind the centre propeller. ABB, which is a large supplier of electro motors has had several deliveries of its "Azipod" system. This is an azimuth thruster where the electromotor is located down in the propeller housing itself (Figure 5.50). In the propeller system delivered by Rolls-Royce (Aquamaster and similar) the electromotor will be located inboard and will transfer power via a mechanical system to the propeller (Figure 5.51).

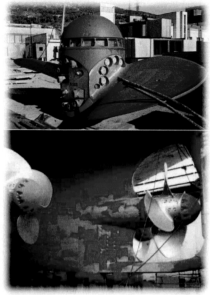

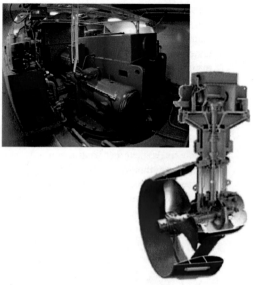

Figure 5.51
Rolls-Royce azimuth propeller (Aquamaster) where the motor is located in the ship. The photo shows the electromotor in front of the gearbox.

Figure 5.49
The propellers on the Russian nuclear icebreakers. Above is shown a photo of a propeller with two of its blades dismantled.

Figure 5.50
ABB Azipod. The photo above shows the inboard part of such thruster on the KV Svalbard.

Diesel / Mechanical propulsion
Because of the relatively large loss with D/E propulsion, on some icebreakers diesel-mechanical (D/M) propulsion is used. Examples of this is the Swedish *Oden*, the Russian *Madyug* class, the Norwegian/Swedish *Viking* class, and a couple of new Canadian icebreakers. This has been possible through reducing the danger of ice in the propeller, amongst other things with the aid of modern design of stern and propeller nozzle. The propeller nozzle contributes in addition to increase bollard pull. With such propulsion it will be essential to use a variable pitch propeller in order to maintain flexibility during manoeuvring. The danger of great and rapidly varying loads on the engine is eliminated by coupling large flywheels between the engine and gears (Figure 5.52).

Variation and load as a result of sailing in ice, as well as the relatively extreme manoeuvres could result in large temperature swings and fatigue of the materials in the engine. Turbochargers are also relatively exposed to damage under such loads. On *Oden* the powerful dimensioning has meant that there have not been any problems with the engine and the mechanical propulsion. On the Viking ships, which, not to the same degree as *Oden,* are built for extreme loads, several types of damage have occurred during operations in ice. There has been turbo breakdown and damage to propeller blades (Figure 5.53). After the 2003 season, all eight propeller blades on the *Vidar Viking* had to be changed as a result of ice damage.

Figure 5.52
An extra large flywheel can absorb the largest variations in loads imposed by ice and reduce engine wear (photo of Oden)

Figure 5.53
Ice damage on propeller blades on the Vidar Viking. The ship is a Moss-808 Ice design (Photo: A. Kjøl).

Nuclear propulsion

Nuclear power is a very controversial form of energy which at the end of the 1950s was thought would revolutionise shipping. The icebreaker *Lenin* which was launched in 1959 was the first civilian surface ship to be powered in this manner. The big advantage of nuclear power on icebreakers is, as previously mentioned, the large range and operating time the ships have. With a supply of a few hundred kilos of enriched uranium, the ships can operate for 4 years without the need for energy supply. An equivalent powerful diesel icebreaker will during the same period have consumed about 100 000 tons of oil, and had a requirement for bunkering many times. Diesel icebreakers will have a need for a bunkering installation in the Arctic (Kjerstad,1990). This characteristic is of significance for safety and nuclear supporters will assert that it is also the most environmentally friendly manner in which to operate ships in the Arctic. Opponents are of the opinion that nuclear power is objectionable and direct special focus on the waste problem and the danger of radiation in the event of accidents (Figure 5.54).

Figure 5.54
The nuclear shallow water icebreaker Vaygach. The engine is 44,000HK and is coupled to 3 propellers.

As previously mentioned, Russia now (2011) operates 6 nuclear-powered icebreakers in the Arctic (+3 taken out of service), as well as a nuclear-powered LASH carrier which is considered to being modified as a drilling ship. Some of the ships are fairly old and will in all probability be taken out of service during the next ten years. Canada has also considered nuclear power for its planned Polar-8 ships. Canada has rejected these plans for financial reasons, while Japan has built an experimental vessel based on nuclear power. Therefore, a short description follows of this type of propulsion, without getting into reactor technical details. In principle, this method of propulsion is an advanced edition of what we know from steam turbine ships, i.e. older tankers and navy vessels. As is shown in Figure 5.55, the reactor functions as a"boiler" and generates steam for the turbine. The turbine could in principle well be coupled directly to the propeller, but this would be a bad solution on icebreakers. It is therefore usual that the turbine drives generators which are then coupled to electric motors. By using electro motors the manoeuvring will be very flexible and provides opportunities to use several propellers.

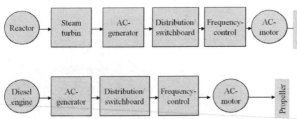

Figure 5.55
Block diagram of the engine in a modern nuclear powered icebreaker (above) compared to a ship with D/E propulsion (below).

From Figure 5.54 we see an arrangement on the nuclear icebreaker class *Taymyr*. The reactor compartment is well protected in the centre of the ship and propulsion is done by 3 propellers. The ship is designed to tolerate

collisions with ice-going merchant ships (SA-15) and the safety systems are very extensive. Amongst other things, there are "back-up" systems for the reactor cooling and the ventilation installation closes automatically in the event of any leaks and radioactive gasses. Even though oil is not used for powering, the total costs of operating such ships will be very high. The complexity means that large crews with special competence are required. (Figure 5.56).

Figure 5.56
The control room on a nuclear icebreaker is manned continuously by 5-6 persons. Here from the control room on Russian Yamal.

Cooling systems and tanks

There are especially strict requirements for all ice-classed ships regarding the cooling system on the engines. The reason for this is that the intakes for the cooling water can be blocked by ice. In such cases, the engine will overheat quickly and there could be a"black-out". The alarm can also be triggered as a result of a fall in pressure of the cooling water. In order to avoid this, there will be several relatively high and large intakes ("sea chests") from which the water can be taken (Figure 5.57). Under the intake there will be a coarse grid, while inside there can be a finer, tilted grid to prevent ice blocking the pipe system. In order to keep the intake free of ice, the cooling water can be directed back to the intake. A lot of ships, under normal operation, will have the automation to direct some back and some overboard. If the intake gets blocked, there will also be the possibility of circulating cooling water in ballast tanks, but with full power, the time that can be done will be limited to a few hours. It is also essential to have "heating" in the tank system to prevent freezing, and heat from the waste heat recovery unit could be used. In addition, water will normally be circulated in the fresh water tanks.

Operation of ships in extremely cold regions could also lead to problems with wax formation in the fuel tanks and fuel system. It is therefore important that "winter quality" fuel is used.

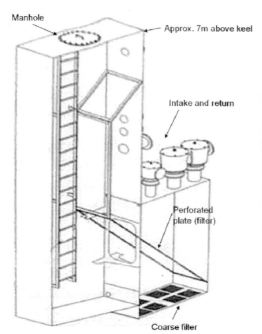

Figure 5.57
Intakes (sea chests) on ice-classed ships are specially designed to prevent ice blocking the cooling water to the engine. Here is a drawing of one of the two largest intakes on the KV Svalbard. In addition, the ship has 5 smaller intakes.

Protection and dimensioning of propellers

A ship with an ice class will always have a need for an extra reinforced propeller. There will also often be measures to reduce the danger of ice clogging the propeller. In general one can say that it is desirable to get the propeller as deep as possible, which will also contribute to increasing the bollard pull. Figure 5.58 shows the design of the stern of the research vessel the *Ernest Shackleton (previously the Polar Queen)*. The ship has D/M propulsion and is classed as Icebreaker Ice-05 in DNV. Note the deflecting fins that are mounted in front of the propeller to lead the ice away from the propeller. By designing the stern and propeller arrangement in this manner, the amount of ice that comes into contact with the propeller will be too small to do any damage. The danger of loads on the propellers, however, could arise when reversing and manoeuvring in the ice.

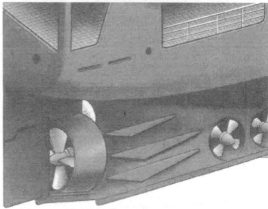

Figure 5.58
The stern of the "Polar Queen". Note the deflecting fins in front of the propeller.

Since propeller damage is amongst the most usual damage that can be done to an ice-going vessel, the loads which can be imposed on a propeller are described. The description is based on a semi-empirical model which is used on Russian icebreakers (Ignatjev) (Figure 5.59). Huge loads arise when the ice passes through the propeller, or gets stuck between the propeller and the hull. This arises most often when the propeller contra-rotates to the direction of speed. In such situations the propeller will be slowed down considerably and will often stop completely. If the propeller is at a standstill and the ship continues to make way, it is important to be aware that the propeller blades can be exposed to huge loads and therefore be bent. The propeller must therefore never be stopped when the ship makes way.

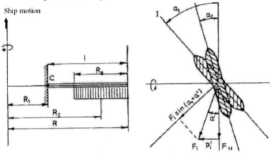

Figure 5.59
The ice load and resulting force on a propeller blade (Ignatjev).

It is estimated that the load is evenly distributed over the outer 2/3 of the blade (l), and that the resulting force (F_i) creates an angle, α', with the propeller blade. The maximum bending moment (M), on the weakest point (C) will therefore be:

$$M = F_i * (R_2 - R_1)* \sin(\alpha_1 + \alpha')$$

M = max. bending moment at the root of the propeller blade (at the hub).
F_i= resulting ice load on the blade.
R_1= radius of the propeller hub.
R_2= radius of the centre of the ice load.
α_1= the blade's angle of pitch.
α'= angle of attack on the resulting force.

By only estimating the force in the propeller blade's plane, we have:

$$F_{ti} = \frac{M_{i\ max}}{R_2}$$

$M_{i\ max}$ = maximum moment on the propeller with ice loads.

The maximum ice moment is determined by full scale measurements, and Ignatjev presents them as a linear connection with the propeller diameter. Another model (Enkvist and Johansson) are of another opinion that the quadratic connection is more realistic (Figure 5.60).

In order to be able to provide a suitable estimate for the sectional modulus (W) on the wing, because of ice loads, Ignatjev employs a safety factor of 3.

$$W = \frac{3 \bullet M}{\sigma_y}$$

W = sectional modulus on the blade.
M = bending moment due to ice.
σ_y= tensile strength of the propeller materials.

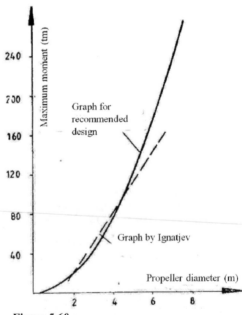

Figure 5.60
The construction moment for the propellers on ice-going vessels (Enkvist and Johansson vs. Ignatjev).

Icebreaker, propeller	Diameter (m)	Number of blades	Pitch-ratio	W/W_e	Number of blades
Lenin, wing	4.8	4	0.70	1.21	0
Moskva, wing	4.82	4	0.63	0.65	6
Moskva, centre	5.8	4	0.50	0.70	2
Kiev, wing	4.82	4	0.77	0.97	0
Severnyi polpic	5.13	3	0.70	0.24	15
Sibir	4.40	4	0.92	0.61	2
Lena	4.60	4	0.59	1.36	0

Table 5.4
Damage that has occurred on some of the
older Russian icebreakers. Note the
connection between the sectional modulus
and damage to the propeller. W = real
sectional modulus, W_e = estimated
sectional modulus.

In Table 5.4 we can see the calculation of
the sectional modulus as a function of
damage which has occurred on some older
Russian icebreakers. From this we can see
that the old icebreakers *Kiev, Lenin* and
Lena, where the sectional modulus is greater
than or similar to Ignatjevs proposal, no
cracks on any of the blades have been
detected. Based on the model tests, it has
been concluded that the design of the
propellers of ice-going vessels should be
different from what is used on ordinary
ships. The tests show that high pitch
conditions and small diameter gives the
most favourable combination in ice. In order
to achieve satisfactory strength the
icebreaker propellers are also characterised
by thicker hubs and blades (Figure 5.49).

Swedish State icebreakers which are used in
the Gulf of Bothnia have not had any
problems with ice damage to the propellers.
Ymer (*Frej* class), however, had problems
with bow thrusters on a mission in the
Arctic. The hired Viking ships have also had
some propeller damage in difficult ice
conditions in these waters (Figure. 5.53).

The materials which are used in propellers
of ice-classed ships are normally NiAl
bronze or stainless steel. The tensile strength
of stainless steel is 0.2% higher than NiAl
bronze. Blades manufactured from stainless

steel are also normally
thicker. Based on this,
it could be believed that
stainless propellers
were to be preferred.
An extensive
examination of ice
damage to propellers
performed by Det
Norske Veritas, however, cannot establish
any marked difference regarding the two
types. When damage has already occurred, it
can often be possible to replace the blades
that are damaged, since even the fixed pitch
propellers normally have replaceable blades
(Figure 5.61).

Figure 5.61
The Russian SA-15 ship Kapitan Danilkin
collided with ice frozen to the bottom on
the Siberian coast. Here we see the damage
to the propeller blades which had to be
replaced.

The large loads which arise when ice penetrates the propeller (Figure 5.62) will naturally be transmitted to the shaft, and it often happened in the infancy of icebreaking that the shafts broke. This occurs very seldom today, but it is not unusual that the shaft bearings are damaged. If there is damage to the bearings in an oil-filled casing, this can lead to two problems:

- Overheating and operational problems.

- Oil leaks with subsequent damage to the sensitive Arctic environment. This is very serious both for the environment and the ship which can be faced with large amounts of compensation. It is not unusual that azimuth thrusters and housings leak considerable amounts of oil due to large ice loads.

In order to reduce the dangers of pollution, in recent years the focus has been on water-lubricated and laminated bearings of synthetic materials.

Figure 5.62
On the ice floe studied here one can clearly see traces of the propeller. The propellers on icebreakers must be dimensioned to tolerate such large loads (Source: Aker Arctic).

5.2.4 Rudder, manoeuvring organs and heeling system

Rudders on ice-going vessels are very exposed to large loads. Over the years this has resulted in numerous bent rudder shafts. The risk of damage can first and foremost be reduced by correct operation and powerful dimensioning. On many of the new vessels with D/E propulsion and azimuth thrusters, one will naturally have eliminated the necessity for a rudder, even though there are also vertical constructions which under extreme conditions can be damaged by the pressure of ice.

As a curiosity it can be mentioned here that Colin Archer, as early as with the construction of the polar ship *Fram*, in 1892-93, understood the significance of protecting rudder and propeller. The rudder and propeller were constructed so that one could draw them up directly through a chute in the stern of the ship in case the ship was exposed to pack ice. This was probably one of the reasons that Nansen's operations over the Arctic Ocean (1893-96) were a success.

On the operational side it is a requirement that it is ensured that the rudder is in the centre position when reversing. Ice which is pressed under the stern will then come into contact with the fin which is located behind the rudder, and the chances of damaging the rudder shaft are reduced. On some newer icebreakers, such as the *Oden* and the Viking class (Moss-808 Ice design), the ship's two rudders are used as a protection for the propellers while reversing. The rudders are then automatically placed against one another, and because the rudders rest against blocks, damage to the rudders is avoided (Figure 5.27).

The class requirements when dimensioning the rudder shaft, bearings and bushings, are mainly based on the loads which are a multiple of the pure hydrodynamic force and torsional moment. For most ice classifications a factor of 2-3 for the hydrodynamic force, and 5-7 for the torsional moment, can be used. In order to give a picture of the diameter of the rudder shaft (d) the empirical formula developed by Faddeyev can be used:

$$d_r = 100 \bullet \sqrt[3]{\frac{K \bullet A_r \bullet r}{\sigma_y}}$$

d_r = diameter of the rudder shaft (cm).
A_r = rudder area (m).
r = distance, the centre of gravity of the rudder to shaft (m).
σ_y = tensile strength, shaft material (kp/cm).
K = empirical constant.
K = 25 (SHP 2,000).
K = 20 (12,000-25,000).
K = 15 (6,000-12,000).

The rudder area, A_r, according to Russian recommendation, is given the following criteria:
A_r = 0.016* L * T (triple installation)
A_r = 0.024* L * T (double installation)
L = ship's length
T = ship's draught.

In comparison it can be mentioned that the rudder area of the *Oden* is 23 m², which is somewhat larger than that used for Russian icebreakers.

In order to improve the turning characteristics, the heeling system is often used in addition to the rudder. From the turning circle data on the Russian icebreaker *Madyug* (Figure 5.63) we see that by using a 4.5 degree heel, a reduction of the turning circle of 27% can be achieved. Figure 5.63 also shows the effects of the thickness of the ice on the ship's turning circle. On ships with a "turning-reamer" (ex. *Oden*) this effect will be even more pronounced.

A problem for all icebreakers is rolling in open sea. The reason for this is that the hull

often has very rounded ship sides, which gives little residual stability (waterline area / surface inertia increases only little when heeling). In addition to this, for operational reasons it is not possible to use a bilge keel. Stabiliser fins and passive stabiliser tanks will also be unsuitable. However, there is a system which can be effective in open waters. By connecting a special heeling sensor and calculating unit to the autopilot the rudder can be used to give the ship a heeling moment which is in contra-phase with the rolling movement. Similar possibilities will be on modern tugboats with two Voith-Schneider thrusters, but these will normally not be suitable for ice-going vessels.

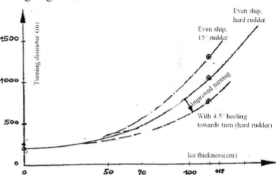

Figure 5.63
The thickness of the ice and heeling influence on the turning circle of the icebreaker Madyug.

Galvanic protection
All ships which are built in steel require a form of protection against galvanic pitting. The usual is that sacrificial anodes of zinc are installed on the hull, rudder and housing. On ice-going ships it will be required that they are not exposed to ice which can break them loose. An alternative that is used on some icebreakers is an actively imposed tension instead of sacrificial anodes.

5.3 Merchant ships

When one talks about merchant ships for arctic regions, it is normally about ships with a so-called Baltic ice class in DNV:

1A*F, 1A*, 1A. The classes 1B and 1C can be used early in the winter, i.e. in the Gulf of Bothnia. In Norway, the Gulf Stream has contributed to almost eliminate the necessity for ice-classed vessels in coastal areas off Svalbad (and some years the Oslo fjord). For this reason, there are few merchant ships in Norway with a high ice class – to date (2011) there is only one ship over 10,000 tons in the categories 1A* or 1A (Express Coastal ship *Fram*). It is first and foremost Russia, Canada, Sweden and Finland which have built conventional merchant ships with a high ice class. In Norway and Denmark there are mainly some fishing vessels, offshore vessels and smaller expeditionary ships with the equivalent class.

5.3.1 Description of some merchant ships

Norilsk Class / SA-15 (Russia)
This type of vessel can be described as specially constructed to sail the Northern Sea Route (Northeast Passage). The designation SA stands for Sub-Arctic and 15 stands for the ice thickness which the ship is designed to handle (1.5m). The first ship of this class was launched in Finland in 1982 (Figure 5.64). The ship has diesel-mechanical propulsion with two engines, each of 10,500hp coupled to a relatively large variable pitch propeller. The engines can be operated individually or together. Often one engine is used in open sea and two in ice. Up to 1987, 19 such ships were built – most of them for the Murmansk Shipping Company, which operates icebreakers and a long range of other ice-classed ships (http://www.msco.ru/).

Equipment can vary somewhat, and some ships are also more ice reinforced than others. The ships have regularly sailed the length of the Northern Sea Route. In 1985 experimental voyages were started with SA-

15 ships. The trips went from Vancouver to Arkhangelsk in November-December and were an attempt to extend the season for the Northern Sea Route. This type of ship has proved to be well suited to the traffic on the Siberian coast. Figure 5.65 shows the ships' data and general arrangement.

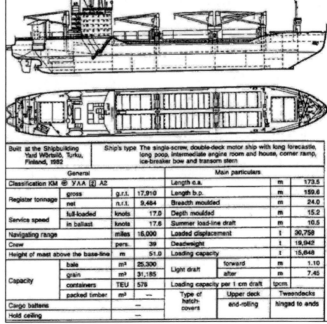

Built at the Shipbuilding Yard Wärtsilö, Turku, Finland, 1982			Ship's type The single-screw, double-deck motor ship with long forecastle, long poop, intermediate engine room and house, corner ramp, ice-breaker bow and transom stern		
General			**Main particulars**		
Classification KM ⊕ УАА ⬜ A2			Length o.a.	m	173.5
Register tonnage	gross	g.r.t.	17,910	Length b.p.	m	159.6
	net	n.r.t.	9,484	Breadth moulded	m	24.0
Service speed	full-loaded	knots	17.0	Depth moulded	m	15.2
	in ballast	knots	17.6	Summer load-line draft	m	10.5
Navigating range		miles	16,000	Loaded displacement	t	30,758
Crew		pers.	39	Deadweight	t	19,942
Height of mast above the base-line		m	51.0	Loading capacity	t	15,648
Capacity	bale	m³	25,300	Light draft	forward	m	1.10
	grain	m³	31,185		after	m	7.45
	containers	TEU	576	Loading capacity per 1 cm draft	tpcm	
	packed timber	m³	—	Type of hatch-covers	Upper deck	Tweendecks
Cargo battens			—		end-rolling	hinged to ends
Hold ceiling			—			

Figure 5.64
Data and general arrangement for SA-15 ships which are used on the Northern Sea Route.

We note especially some details on the hull from the general arrangement. The subsea hull is designed to tackle difficult ice conditions. The icebreaking bow and convex ship sides under the waterline (Figure 5.65). The thickness of the plating in the ice belt and bow is 36 mm. Loading and unloading can take place via the ramp at the stern. This will make operations easier on the ice. The stern is designed for connection / towing of vessels which are in convoy (towing notch). Air jets are installed along the sides under the waterline. Heated air can be blown out and will considerably reduce friction. A command centre is installed in the forward mast which can be used in convoy and during towing. However, Russian seamen asserted that this is used very little.

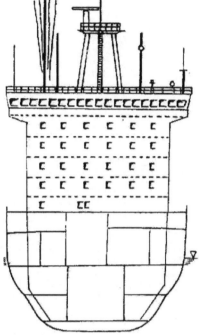

Figure 5.65
Subsea hull on the SA-15 ships have convex ship sides (flare). The ship will therefore be able to resist difficult ice conditions and pack ice.

If we look at the class designation which is indicated, it is in Russian, and has the following notation:
KM * YΛA 2   A2

KM*= Ship with own engines which are classified.

YΛA= Highest ice class for the Arctic (e.g. icebreakers)

2 =   Two adjacent compartments can be filled without the ship sinking.

A2 =   Unmanned engine room, manned control room.

The ice class will approximately equate to Ice-1A* in DNV.

Norilskiy Nickel class (Russia)

This ship is a prototype of a series of five ships which is intended to replace the SA-15 ships. The first ship was commissioned in 2006 (Figure 5.66), and will mainly sail between Murmansk and Dudinka on the Yenisei river, where the ore from Norilsk is shipped out. The ship has D/E propulsion with an azimuth propeller (Azipod) and is built according to the "Double-Acting" principle. Under normal ice conditions they will operate without icebreakers. The principle dimensions are:

Length:	169.5m
Width :	23.1m
Draught:	9.0m
Engine:	13,000kW (17,400hp)
Icebreaking capacityt:	1.5m
Dead weight (load cap):	14,500t
Containers:	650 TEU
Gross tonnage:	approx. 16,000t
Ice class (RR):	LU7 (λy7)

Arctic Container Vessel
NORILSKIY NICKEL   Aker Arctic

Figure 5.66
Norilskiy Nickel is the ship type which is the main successor to the SA-15 ships for transport on the Northern Sea Route in Siberia (Source: Aker Arctic).

Experiences with the ship type are very good and there is reason to believe that this will form the school for future shipping in the area.

LASH ship *Sevmorput* (Russia).
In 1986 the 61,000 tonn LASH ship "Sevmorput" was built in Russia (Figure 5.67). The ship is nuclear powered and specially built to sail the Northern Sea Route to the large river estuaries in the east. The principal dimensions of the ship are:

Length/width:	260 m / 32.2 m
Draught load/empty:	11.7 m / 4.4 m
Deadweight:	31,900 ton
Container cap. :	1,336 TEU
Engine power (nucl.):	approx. 40,000hp
Speed:	20.7 knots
Range:	Unlimited
Ice class:	YΛA

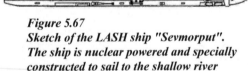

Figure 5.67
Sketch of the LASH ship "Sevmorput".
The ship is nuclear powered and specially
constructed to sail to the shallow river
estuaries in Siberia.

At present (2011) it is uncertain whether, or when, more merchant ships of this type will be built. The development will probably be dependent on much of the future attitudes regarding nuclear energy. At present there are large limitations regarding harbours which can be visited with nuclear powered vessels. This applies both in and outside Russia. The Russian shipping companies do

not hide the problems this entails in the operation of nuclear powered icebreakers and merchant ships. The operation of the ship is also extremely expensive and the lightering possibilities are unusable as soon as the rivers freeze over. It is therefore now (2011) evaluated to modify the ship as a drilling ship for operation in Arctic regions.

Bulk carrier *Arctic* (Canada).
This OBO / bulk carrier came into operation in the Canadian Arctic in 1978, and was then a modern double hull ship which satisfied the "Class 2" Regulations in Canada. During winter navigation the ship received some damage and it was decided to modify / reinforce the ship (Figure 5.68). In addition to the hull reinforcement, an advanced navigation / routing system was implemented. The ship is operated by Fednav which has several other ships with a high ice class (http://www.fednav.com). The ship has the following dimensions:

Length:	197 m.
Width:	22.9 m.
Draught:	10.9 m.
Dead weight:	28,860 tons
Engine:	11,014 kW (14 780hp)
Speed:	14 knots (open waters)
	1.8 knots (1 m. level ice)

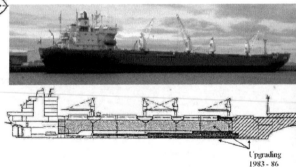

Upgrading
1983 - 86

Figure 5.68
The sketch shows the bulk carrier "Arctic"
after upgrading of the hull.

Oil tanker *Stena Arctica* (Sweden)
The ship was built in 2005 in accordance with Ice Class 1A-super and is thereby suitable for year-round navigation in Swedish and Arctic waters (Figure 5.69). The high ice class means that the ship has approximately 10% more steel, and 65% stronger engine power in relation to conventional ships. The Stena shipping company also operates a range of other ships with a high ice class (http://www.stenabulk.com). The ship type is of the so-called Aframax type, and has the following dimensions:

Length: 250m
Width: 44m
Draught: 15.4m
Dead weight: 117,099t
Engine power: 16.6 MW (22,300hp)
Consumption: Approx. 50t / day at 14 knots
DNV Class is: 1A1, Tanker for oil, ESP, ICE-1A*, E0, NAUT-AW, VCS-2, PLUS-1 TMON, NAUTICUS

Figure 5.69
Stena Arctica is the largest ship with Ice Class 1A- super that has been built.

Oil tanker *Vasily Dinkov* (Russia)
The ship is designed by Aker Arctic as a Double Acting Tanker with two Azipod propellers and is built by Samsung in Korea (Figure 5.70). The ship was commissioned in 2008 as the first of a series of three which are intended used for shipment of oil from Northwest Russia. The ship has a high ice class and is expected to be able to operate in 1.7m thick ice. The price was USD 168

million. The operator of the ship is Sovcomplot (www.scf-group.com) and the dimensions are as follows:

Length: 256m
Width: 34m
Depth: 14m
Engine: 2 x 10MW (26,800hp)
Ice class: LU 6
Dead weight: 70,000t
The class in the Russian register is: RMRS KM +LU6, +1A1 "Oil tanker" (ESP) with exclusive approval for shaft power ABS +A1(E), Oil Carrier, SH, SHCM, +AMS, +ACCU, VEC, SPM, MIBS, ESP, Baltic Ice Class Equivalent to RS Ice Class LU6 Ice Breaking Tanker.

Figure 5.70
The oil tanker Vasily Dinkov is the first of a series of 70,000 tonners that are built for shipment from Northwest Russia.

119

5.4 Questions from Chapter 5

1)
What separates the "Baltic" class requirements from ice classes for operation in Arctic regions?

2)
What ice conditions are ships with DNV Class Ice 1C made to operate in?

3)
What IACS class must a ship have if it is to operate year-round in all arctic waters?

4)
What ice class must a ship with Germanicher Lloyds class have if it is to sail to the Gulf of Bothnia where there are requirements for Ice Class 1B?

5)
How thick must the plating be for a ship with ice class Ice 1A (DNV)?

6)
Name some disadvantages which an ice classed ship has in relation to convention ships.

7)
Name some important operational characteristics of an icebreaker.

8)
What is the mechanical reason that an icebreaker bow should come into contact with the ice at a pointed angle?

9)
What is meant by "close-couple towing"?

10)
What can be the reason that Russia does not allow use of bulbous bows on its Northern Sea Route?

11)
How are rudders on ice-going vessels normally protected against ice loads?

12)
Explain what is meant by "Double Acting" design.

13)
Why is it effective to flush water in front of the bow of a ship which is sailing in heavy ice?

14)
What is meant by "duck-walk" and why is it used on some icebreakers?

15)
What can the reason be that ships with azipod travel faster astern than forward in ice?

16)
What is the reason that the engines of ships that sail in ice often overheat and how can this be avoided?

17)
Name the disadvantages of diesel-mechanical propulsion of ice-going vessels.

18)
Why is it an advantage to have convex ship sides on ships that are to operate in heavy ice?

6 Operation of ships in arctic regions

In this chapter, the general problems connected to operation of ships in ice will be described. It can therefore be adapted to most of the ice-covered waters. Nevertheless, it can be worth noting that several organizations and countries have issued operational descriptions in order to safeguard sailing in cold and ice-covered regions. Examples of such sources are BIMCO, OCIMF, Swedish Maritime Authorities and the Canadian Coastguard.

6.1 Ship handling

6.1.1 Navigation in ice-covered waters

Being able to navigate a ship in ice is not something one can do from books. Practical experience and a good knowledge of the ship's performance in ice and the static and dynamic characteristics of the ice are required. For people who do not have this experience it can be useful to summarise some general experiences and thereby possibly make the learning time shorter.

There are a series of preparations which should be made before one travels into waters with ice and often extreme cold. Important points can be:
Operational preparations in the form of planning. Here it is important to have obtained statistical data and forecasts regarding ice conditions, as well as information regarding the applicable rules for the area. One should also see that the necessary procedures are reviewed, and possibly posted on the bridge. Lookout and other preparedness should be intensified and strengthened.
Technical preparations such as draining water in pipe systems that can freeze. Included in this is also draining of ballast water, so that one is certain that there is no water in the ventilation and observation tube. Seeing that the rescue materiel is suitable for the conditions – in this is

especially included seeing that the lifeboat engines have fuel which is intended for the cold temperatures that can occur.
Useful equipment which one should evaluate installing (or checking the functionality of) can be: Ice searchlights, night binoculars, thermometer for the sea water on the bridge, equipment for reading of ice charts, etc.

When sailing in ice, the vessel must be propelled forwards in the most careful way without deviating too much from the main course. In general, vessels without ice reinforcement must not travel alone into drift ice which covers more than 4/10 of the surface of the sea (ref. Chap. 4), even though it is only a limited ice belt.
Remember – the quickest way through the ice is seldom the shortest! (Figure 6.1).

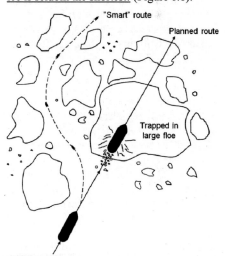

Figure 6.1
The quickest way is seldom the shortest (CCG, 1992).

The best conditions for navigation in drift ice are days with clear air and an even, thin cloud base. When one approaches the ice, one will be able to see "ice blink" in the sky in the direction of the ice. The course should normally be steered towards where "the water sky" stretches furthest forward. Channels will be seen as dark strips in the ice blink (Figure 6.2).

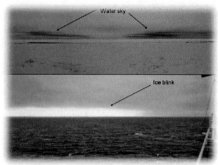

Figure 6.2
The ice conditions at the horizon can often be interpreted by the reflection in the cloud base. Here the dark "water sky", seen from an area with dense ice and "ice blink" seen from open waters.

Where the ice blink is weakest, the probability for <u>open</u> drift ice will be greatest. This effect will be best visible after a snowfall. If there is a cloudless sky there is no possibility for normal ice blink. If there is a swell in the ice, this is often a sign of lighter ice conditions from where the swell is coming. A yellow-white fog, however, can mean the presence of ice. Ice edges far away also have a tendency to tower up and appear a lot thicker than they really are. Under such conditions, openings in the ice will appear as dark strips. In reduced visibility one must be on the lookout for other advance warnings of ice. If the wind is not directly on the ice edge there will often be small ice lumps in the water. A sudden fall in temperature, at the same time as the water temperature moves towards zero or lower, will also be an advance warning. For such observations it is very useful to have a thermometer for both air and water in a place on the bridge where they are easily seen. At the ice edge, the swell also has a tendency to subside, and the wind will blow more evenly. In other words, it is very important to keep a <u>good lookout in order to avoid unpleasant surprises.</u> On some ships there will be a separate crow's nest so that one can get as high as possible, and thereby have the best possible view of the ice (Figure 6.3).

Figure 6.3
From the crow's nest (above) the navigator will have a considerably better view of the ice conditions. Here from the KV Svalbard.

In calm weather with fog, the breakers can be heard over long distances. Very near the ice, the fog will be white and milky, and one will normally discover the ice at a sufficient distance if speed is reduced. In snowy weather and wind there will be larger problems with discovering ice. In such conditions the radar will also have large limitations, which will be described in the next chapter. Under such conditions, one should evaluate heaving to, pending better visibility.

When one sails into ice, one must study it carefully in order to find the most favourable position of attack. It is assumed then that the navigators have obtained all information possible about the ice conditions, in the form of charts and satellite imagery. It can often pay to sail a little along the ice edge and search with radar or visually, to find openings. If the wind is blowing on to the ice edge, it can be expected that the ice conditions are lighter further in (Figure 6.4).

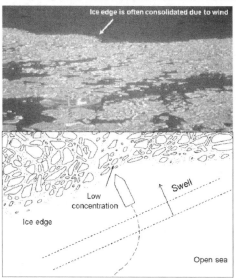

Figure 6.4
When the wind is blowing on to the ice edge, the ice can often be somewhat lighter further in. Enter with the swell astern.

In the opposite case, with the wind from the ice, the ice will usually be denser further in. If the ice edge consists of spread drift ice, there will be swell far into the ice. In such cases it is important to enter the ice edge in the same direction as the swell to avoid damage to rudder and propeller (Figure. 6.4). This concerns especially smaller vessels. In large floes and heavier ice it will be necessary to break a way through (ramming). Ramming is often done by moving at full speed in to the ice and reduce engine power just before the collision. By reducing the engines, the danger of getting stuck is avoided. By ramming, the ship will slide up on the ice which normally breaks up because of the weight of the ship. If the ice does not break up, the ship will either slide off or get stuck. The loads connected to such operations are described in Chapter 5. It is worth noting that ships with ordinary ice reinforcement (Baltic classes / Type ships) are not built for ramming.

Boring and twisting is as a rule the most effective method of getting through drift ice. By hard rudder the ship can be used as a lever and twist larger ice floes away (Figure 6.5). When the ice opens up it is easy for the ship's speed to increase. In such cases it is important to reduce the speed again, as well as being on guard for especially hard and damaging ice (growlers etc.). In this connection it is important to realize the fact that one can easily be pushed sideways after colliding with ice in a channel. This can lead to the ship's sides colliding with ice from a direction the ship is not constructed to absorb large loads from. Many cases of damage, even on strong ships, have occurred because of this.

When sailing in varying drift ice, the conditions can quickly be changed by wind and current which can form pressure in the ice. This will in its turn lead to increased friction and loads on the sides of the ship. Under such conditions, the ship could get stuck (Figure 6.6). In drift ice, the brash can often be somewhat whiter and the swell a little slacker when ice pressure is building up. If one gets stuck in the ice there are several ways in which the ship can be freed. Regardless, one should attempt to use the propeller and rudder actively to keep a certain circulation and movement along the ship sides.

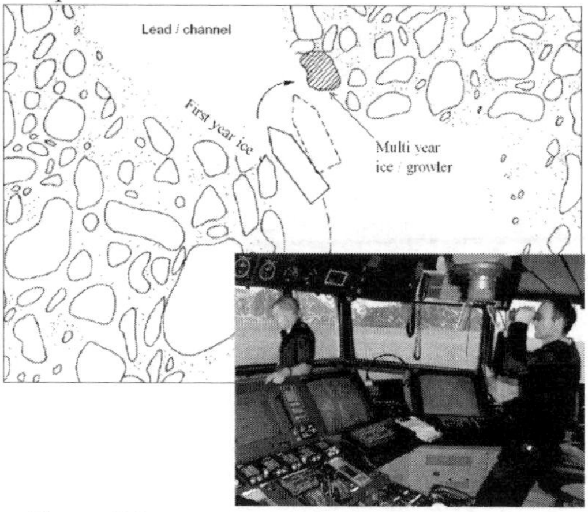

Figure 6.5
When sailing in drift ice, one should twist forward in the channel, at the same time as one regulates the speed and keep lookout for hard and dangerous ice.

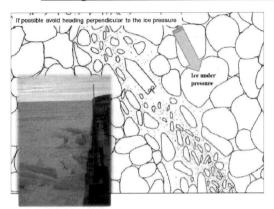

Figure 6.6
Sailing in dense drift ice under pressure can often lead to the ship getting stuck. The pressure on the sides can be extreme.

By trimming and heeling one can also manage to free the ship and thereby get underway. If the ship has no tanks that are specially suited to this purpose, a heavy object can be hung on the boom and it is swung from side to side. One can also cast ice anchors and try to heave the ship free. On hunting vessels it has also been usual to use explosives to clear channels or free the ship from the ice. The charges should preferably be placed under the ice, and so far away that the ship will not be damaged. It is worth noting that the pressure of the ice often lightens at high tide. The tide tables are therefore a useful tool in ice-covered waters.

Reversing
If the ship is to enter into drift ice stern first, the rudder and propeller on conventional ships will be particularly exposed to damage. Reversing must therefore take place with great caution. One method of opening channels behind the ship can be that there are regular powerful thrusts forward so that the propeller wake opens the channel behind. As soon as the ship again reverses the rudder must be put into centre position to avoid damage (Figure 6.7). If the ship has two azimuth thrusters, these can be used actively for clearing the channel, at the same time as the ship moves slowly astern.

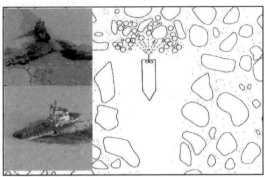

Figure 6.7
Reversing must take place with great care. The propeller wake can be used regularly to open channels behind the ship (Background source: CCG, 1992).

Mooring
Going alongside a quay can often be difficult when there is ice in the harbour. Even though the icebreaker has attempted to clear the area it is seldom ice-free along the quay. Ice that is trapped between the quay and the sides of the ship can also damage both ship and quay severely. Before final mooring is done, therefore, it is important to clear the sides of the ship of ice. An effective method of doing this can be to fix a spring line to the quay and keep moderate thrust forward. This will lead to a stream of water which will gradually suck the ice away from the quay (Figure 6.8).

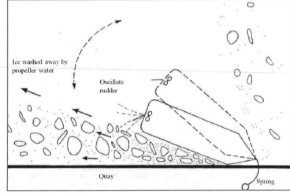

Figure 6.8
Clearing of ice at the quay by attaching a spring line to the bow (CCG, 1992).

When the quay is cleared, the rudder is turned away from the quay such as is usual when the ship approaches the quay. If the ship has azimuth thrusters, these can often be ideal for clearing the ice away from the quay (Figure 4.20). For example, the ship can then lie with its stern towards the quay, with the thrusters abeam in each direction, so that the ship is stationary and flushes the ice away.

6.1.2 Radar navigation in ice-covered waters

As mentioned, in order to obtain an overview of the danger of ice, it is important to obtain information through pilot descriptions and updated ice charts. These can also be important aids in the evaluation of increased preparedness and the intensity of the radar lookout. Lack of lookout and preparedness have resulted in many serious accidents (Figure 6.9).

Figure 6.9
Two damaged ships resulting from a hard collision with ice. The Titanic was a catastrophe, while the Maksim Gorkiy was saved by the Norwegian coastguard.

Detection of ice.
Even though radar is an invaluable aid to the detection of ice, one must always have in mind the large limitations the radar has in this field. The fact that one cannot see ice on the radar is therefore no guarantee

that there is no dangerous drift ice in the vicinity. If one is in an area with smooth, glassy ice, this will not be shown on radar. The strength of the echo from ice will depend on a range of circumstances. Concerning icebergs the slope on the radar-exposed side will mean just as much as the size and distance. In order to detect ice in the best possible way at relatively short distances one should use a powerful 3- cm radar (X-band) with a long antenna, since this has the best detection and reflection characteristics. The pulse length should be medium or long. The radar producer Furuno has also delivered a series of 5 cm radars to vessels that operate a lot in ice, since this combines the characteristics of both the 3-cm and 10-cm radars. However, this 5 cm radar is no longer available since the compromise did not completely fulfill expectations. If one shall detect drift ice or belts of drift ice, growlers (small, hard ice clumps weighing a few tons), etc., the possibilities will be completely dependent on the sea state, height of the antenna and the ice concentration (Figure 6.10).

Figure 6.10
Growlers represent an extreme danger to shipping and they are difficult to detect on radar. Here is an example of a growler in close, but thin drift ice at Svalbard.

Based on experience, one can expect ice detection on radar as follows:

<u>In calm seas</u>
Large icebergs can be detected at 15 - 20
nm (Figure 6.11), while larger growlers
can be detected at 2 nm. Belts of spread
drift ice can be difficult to detect even at
short distances. Dense drift ice belts should
be able to be detected at 5 – 6 nm.

<u>In heavy seas</u>, because of sea echo
(clutter), it will be very difficult, if not
impossible, to detect dangerous ice such as
growlers and belts of dense drift ice. One
must therefore not rely only on radar
observations under such conditions.

Figure 6.11
The radar picture with many echoes of
icebergs. The area is 12 nm (Racal). The
photo above shows the real surroundings.

In order to discover growlers right into the
bow, the distance area should not be over 3
nm. One must also be sure that there is not
a blind sector in the bow which reduces the
performance of the radars. If masts and
constructions create problems with blind
sectors, one should have an antenna placed
in the bow if the ship is sailing in waters
where there is a great danger of drift ice.
So that the echo should not disappear in
the clutter (noise) on the radar screen, the

operator should frequently adjust clutter
control in order to discover possible weak
echoes. The "Echo stretch" function will
be very suitable in open areas with
icebergs and growlers. If one has a radar
with electronic echo trails it can also be
possible to frequently increase gain in
some scans (antenna rotation). A weak
echo from a growler would then appear as
a characteristic stripe in the more random
sea noise (Figure 6.12). Since growlers can
be lumps which have fallen off icebergs,
one should keep a lookout for icebergs
further away.

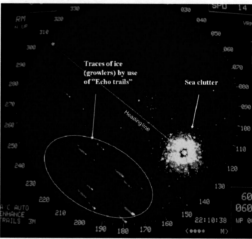

Figure 6.12
Use of echo trails can be useful for
discovering ice which would otherwise be
hidden in sea noise. The photo is of the
Labrador coast.

If the ship is sailing in the ice there will
often be a radar picture which is composed
of various ice conditions. If this picture is
studied carefully, it will be possible to get
a lot of useful information about the
structure of the ice in the vicinity. For
example, with a little experience one can
identify channels, areas with open sea,
pressure ridges or areas with newly frozen
even ice (polynia) (Figures 6.13 and 6.14).
A pressure ridge or areas with pack ice can
clearly be seen at 3 - 4 nm and should be
avoided even by icebreakers and ships
carrying a high ice class. Behind the

echoes from a pressure ridge there will often be a radar shadow that can be misinterpreted to be open waters, which is seldom the case.

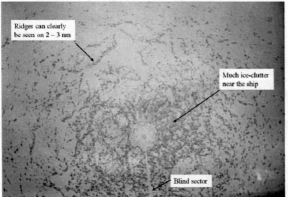

Figure 6.13
The radar picture from the polar pack-ice at the North Pole. The display on the photo is from Norcontrol DB-2000 radar.

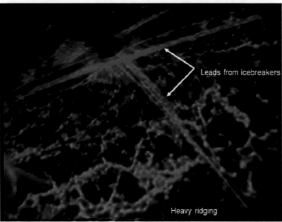

Figure 6.14
The radar picture showing icebreaker channels and pressure ridges in the Gulf of Bothnia.

Behind the larger icebergs the radar shadow can also look like an open channel, but this will normally be straighter at the edges. In the areas with generally even ice one will also be able to identify channels after icebreakers of up to 3 nm distance (Figure 6.14). If the ship is sailing in to a

channel which has previously been broken open by an icebreaker, one must always check if this is navigable with regard to depth, etc. , since it must be expected that ice formations and channels move with the current. There are several examples of ships that have run aground as a result of navigating in an old icebreaker channel.

For the best possible detailed information in an area with ice, a 3-cm radar should be used with the shortest possible pulse length and with narrow beam-width (long antenna). One should also avoid using different automatic noise -reducing functions since they can remove information which can be useful in evaluating ice conditions. Some radars, because of different picture handling techniques, have proved to be completely unusable for interpretation of ice information when one is in ice. In accordance with the IMO specifications, colour separation shall not be used to differentiate between different video levels, but on some radars the possibility is nevertheless found to be able to present data in this manner. When one is in the ice this can give a very informative picture. For example, a ship in the ice can be seen as a characteristic red echo (6.15), while pressure ridges are shown as yellow uneven lines, and perhaps even ice is shown in a light green colour. If the ship is sailing out into open waters again, the radar should be reset to the IMO standard.

Following ships' echoes in the ice can also be useful for monitoring of the convoy behind an icebreaker. In this manner, the officer of the watch on the bridge can adjust the speed and thereby lead the convoy in the most efficient manner. Here, information from AIS can be very valuable (Figure 6.16).

Use of ARPA functions will normally not be possible in ice-covered waters since the echo from the ice will result in the tracker losing the target that is being plotted.

Therefore, one will be able to quickly receive alarms regarding "lost target" or "target swap". A 10-cm radar will normally be most suitable for plotting of vessels in the ice since they will receive weaker echoes from the ice. Since many logs will have problems with showing accurate speeds in ice, it can be difficult to use the radar in true motion. The best alternative will therefore be to use DGPS as speed information to the radar, even though this has its limitations in relation to plotting (ref. Kjerstad, 2010). In ice, therefore, information from AIS is the best basis on which to evaluate the collision danger.

Figure 6.16
AIS provide helpful information of speed in a convoy. Here from Vilkitsky Strait on the Northern Sea Route, August 2010.

The conclusion is that if the ship is in open ice-covered waters, a good visual lookout is important in addition to good radar lookout. A suitable radar range when one is in ice-covered waters can be 3 - 6 nm. In addition, one must be very critical in determine safe speed. The fact that the ship carries an ice class is no guarantee for safe sailing. Systematic studies have shown that the higher the ice class the ship carries, the more often serious damage occurs to the ship – often because of too great a speed and too much faith in that the ice class (strength of the hull) is a guarantee for tolerating all ice loads.

Figure 6.15
Radar and AIS can be useful for following a convoy's movements. Top: The ship's main radar. Bottom: Low radar aft for following a convoy (Photo: A. Kjøl).

Special radar for ice navigation
Throughout the years, many different radars have been introduced that are to have especially good performance for ice detection, as well as classification of ice. Much of this development has taken place in Canada, and especially the system with

differing polarizing of antennas has proved to give promising results. The disadvantage of the systems is that they have been expensive and therefore not had any dissemination among users. Recently, a new system has seen the light of day. This has been developed by Rutter Technology (www.ruttertech.com) and is based on an advanced video processor which is called Sigma S-6. An example in a photo from an ice-covered area is shown at Figure 6.17.

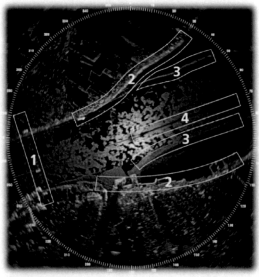

Figure 6.17
Photo from Sigma S-6 radar which is specially developed for operation in ice-covered waters. Here from the St. Lawrence Gulf. 1)Echo of crossing cables 2)Shoreline 3) Fast ice edge 4) Channel behind ships in drift ice.

In connection with research on the use of radar for being able to systematically monitor and classify ice, side-seeking radar systems have been taken into use in aircraft (SLAR) and satellites (SAR). On ships one has researched combinations of different frequencies and polarization. In order to be able to interpret such information it is important to know as much as possible of how different types of ice reflect the radar

signal. It has been proved possible to separate out the thickness of the ice, age and snow cover from qualified interpretations of the signals.

> **REMEMBER !**
> **Never trust that icebergs and growlers can be detected by radar. Therefore always keep a good lookout and safe speed.**

Radar positioning in coastal ice-covered waters.

As previously described, at the coast radar-bearings and ranges to characteristic targets can be used to determine the position with relatively good accuracy. Perhaps the quickest method of using a bearing and a distance to the same point is that which is most used in practice. In waters where ice is found, this is a method which must be used with the greatest caution. The reason for this is that one cannot always be sure that the coastline (Figure 6.18) from which a bearing is taken on radar is the same as the one shown on the chart. This is because of land-fast ice or drift ice which is pushed up against the shore because of wind not represents the precise coast line. This can draw the radar contour far from shore, with the result that the position which is entered on the chart is set too near land (Figure 6.19). Detection by racon is the most certain method for radar positioning when there is ice along the coastline. The position can then be found by taking a bearing and distance from the racon.

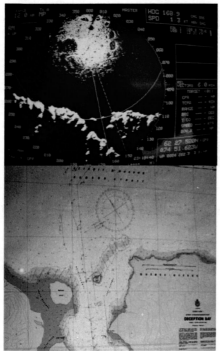

Figure 6.18
Radar picture from the ice at the entrance
to Deception Bay in Canada. Caution
must be shown in using assumed
coastlines on the radar in positioning.

Figure 6.19
Hanseatic aground at Svalbard.
Positioning using radar can be very
uncertain because of ic land-fast ice and
uncertain charts.

Icing on the radar antenna.

On smaller vessels in winter there can often be partially heavy icing from seaspray. In addition, all vessels can be exposed to icing from precipitation and special atmospheric conditions. In such cases, ice can also form on the front of the radar antenna. This will lead to considerable dampening of the transmitted energy, which will mean considerably reduced detection. That ice leads to damping can be understood intuitively by putting an ice cube in a microwave oven (transmits waves at the same frequency range as radar). Almost all the energy which is transmitted in the form of microwaves will be used to heat / melt the ice. If there is ice on the antenna, this must checked frequently and removed if necessary.

REMEMBER !
Ice on antennas and in the rigging can loosen and hit people. Always wear a helmet on deck under such conditions.

Bearing errors at high latitudes.

The meridians close up at high latitudes. For example, 1 nm is one minute of arc at the equator, two minutes of arc at 60° latitude and three minutes of arc at 70.5° latitude. For this reason, errors can arise on the radar display if this is at a very long distance, e.g. 96 and 120nm. If the ship is at 60°N and 135°E and the cursor is set on 62°N and 139E, the real position of the cursor will be a little more to the west. The error is shown on Figure 6.20, and as we see, the error increases between 0 and 90° bearing. The error can in some cases be found in radar manuals, but can also be calculated.

6.1.3 Safe speed, routines and watches.

The International Rules of the Sea of 1977 direct each ship to travel at a safe speed (Rule No. 6). Special emphasis is laid on that in the evaluation of safe speed, consideration must be paid to ice and one must be especially aware of the limitations of the radar with regard to detection of ice. With this as the starting point, many efforts have been made to draw up simple operating instructions which could help the navigator to make the right decisions. Figure 6.21 touches on elements which are included in the determination of safe speed in ice. It will be necessary to go in and analyze each separate element.

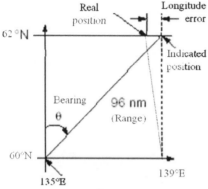

Figure 6.20
Bearing error at long distances can create inaccuracies at high latitudes.

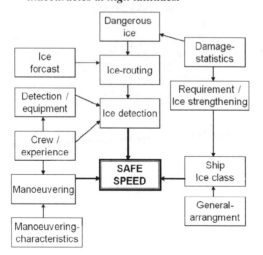

Figure 6.21
Block diagram shows decisive elements for determination of safe speed in ice.

Routines and watches.
When the ship is on its way towards ice-covered regions, it is important that the entire ship's crew is briefed on the situation, and what this entails of special tasks for each one. Both on the bridge and in the engine room the routines will be considerably different in relation to operations in open waters. Sailing in ice is an operation that can be very demanding. To execute the voyage the following routines must be paid special attention:

Operational manoeuvring system.
It is of the greatest importance that the engine is always ready for manoeuvring. This is to ensure that the ship keeps a safe speed in every ice condition that may occur. In the case of ramming it is natural that the engine is always operative. If the ship is sailing in a convoy after an icebreaker, especial vigilance is required since the ship in front can encounter difficult ice and thereby stop. The time available for stopping your own ship is very little. One of the most frequent types of damage that occurs when sailing in ice is an actual collision with the ship in front in convoy (Figure 6.22). Monitoring of the distance between the ships is therefore of extreme importance.

Figure 6.22
If the icebreaker stops because of difficult ice, there can be a great danger for ships in the convoy colliding with the icebreaker.

Manual steering

When sailing in drift ice there will always be a need to deviate from the course to reduce the load from collision with ice floes. In the case of very close ice, one will also have to use the rudder to twist forwards between the ice floes. There will be a need to be able to turn full rudder. In order to achieve the fastest possible turn it is recommended having both pumps on the steering engine in operation.

Positioning / chartwork

As mentioned above, there will often be a requirement for deviation from the main course. This will mean that it can be difficult to keep track of the ship's exact position at all times. If one takes into consideration that most of the Arctic regions are land areas that are at times badly charted (Svalbard, Siberia, Canada, Greenland as well as Antarctica) this can be critical (ref. Chap. 3). One must therefore lay down an unusual amount of work to bring the position up to date. Here one must be especially aware of the danger of sailing in old channels from icebreakers. These can drift with current and wind and several ships have run aground as a result of uncritical navigation in them.

Communication

Communication is especially important in convoy operations. Icebreakers or following ships must be informed about manoeuvres in order to avoid collision, or to avoid that the following ships sail out of the channel and become stuck. Figure 6.23 shows ships in two meeting convoys in the Gulf of Bothnia. In the international signal book (Figure 6.24) will be found a whole series of standardized signals for use between an icebreaker and vessels in the convoy. How far these can be used can vary, and in any circumstances, one must become familiar with the rules for communication in local pilot descriptions.

Lookout with radar and searchlights

In all visibility conditions it is important to keep a sharp lookout in order to choose the easiest route. It can be necessary to combine radar and visual lookout with searchlights, possibly also with night binoculars. IR-cameras (Figure 6.25) have also proved to be well suited for observation of the ice conditions in the vicinity.

Figure 6.23
The Swedish icebreaker Ymer leads a convoy through the ice in Gulf of Bothnia.

Figure 6.24
In the international signal book will be found a series of standardized signals for use between icebreaker and ships in convoy.

132

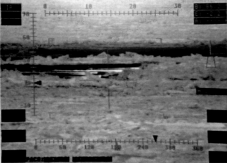

Figure 6.25
Example of an IR photo of dense drift ice.

Tactical planning
If the ship is sailing in convoy it is normally the icebreaker that makes the choice of route. If there is no icebreaker, it is extremely important to choose a route where the most favourable ice conditions can be expected. That will mean interpretation of satellite-based ice forecasts which are transmitted from frequently. These can be both low resolution and high resolution photos / imagery. In addition, a part of the tactical planning can be to interpret meteorological phenomena at the location (temperature and light conditions). In order to tackle all the mentioned tasks a traditional bridge will normally require two navigators and a lookout. This is a solution Russian and Canadian ships always have when navigating in ice. However, in practice there are solutions for completing the voyage with one navigator plus a lookout / orderly. So that this shall be defensible the bridge must be designed taking efficient ice navigation into consideration. Schematically, such bridge can be designed as in Figure 6.26.

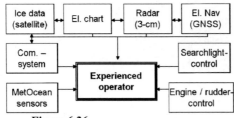

Figure 6.26
The most important elements in an efficient bridge for sailing in ice.

6.1.4 Dangerous ice conditions

Based on an analysis of a series of cases of damage the classification companies have arrived at the conclusion that there are three ice conditions that are especially dangerous for shipping (Fig. 6.27):
a)
Open waters with icebergs, growlers or remains of multi-year ice.
b)
First-year ice with floes of multi-year ice frozen in.
c)
Pack ice consisting of multi-year ice.
d)
Accumulation of ice under ships that sail in very shallow channels.

Figure 6.27
Especially dangerous ice conditions. Top: Open waters with growlers. Middle: First-year ice with growlers or bits from icebergs. Bottom: Close drift ice with some multi-year ice – usually in strong movement.

The reason that multi-year ice is so much more dangerous is that the hardness of it is so much greater. This is due to a lower salt content.

Damage due to category a) can first and foremost be avoided by reducing speed when the conditions for detection become bad because of the dark or reduced visibility (Figure 6.27). This can be a difficult dilemma for a crew which is usually forced to keep to a tight sailing plan. It was such a situation which was the reason for the very dramatic accident to the *Maxim Gorkiy* in 1989 (Figure 6.9 and 6.28) and the *Explorer* in 2007.

Figure 6.28
Damage to ships which have run into ice floes at great speed. Here from provisional sealing of the Maksim Gorkiy (Photo: Norwegian Coastguard).

Conditions of category b) will be far more difficult to determine since the ice can often be covered in snow. A solution to the problem could be to divide the waters into zones which could give guidance as to speed reduction. This is a method which is used in t he Canadian Arctic. The radar-based (SAR) satellites will also be able to detect the age of the ice, and thereby be a good determination tool for navigators on watch. However, this will normally require that the ice data has been interpreted by a specialist.

Voyages in category c) will be a very hazardous matter unless the ship carries the necessary class for sailing in this type of ice (ICE or POLAR in DNV). Extensive local and global loads will also occur, and the

danger of freezing fast with the consequent danger of pack ice will be imminent (Figure 6.29).

In large parts of the Northern Sea Route there is first-year ice which is somewhat less dangerous that the aforementioned categories. North in the Kara Sea and in the Vilkitsky Strait, however, there will be conditions such as described above. Very difficult conditions can also occur when the ice moves out into the large rivers on the Siberian coast. Then for a shorter period the waters cannot be navigated by the large icebreakers. In the rivers themselves there is of course no multi-year ice, but because of the extremely low temperature (-50°C) the ice can be very hard. In some years very difficult conditions can occur in the East Siberian Sea, and this will especially be the case with northerly winds. In most of the Northwest Passage, partly dense and difficult multi-year ice can be expected. In the Gulf of Bothnia and on the Canadian lakes, however, there is never multi-year ice. In the Antarctic, apart from icebergs, there will almost exclusively be one-year ice.

Figure 6.29
Top: Light damage to the side of the ship from ice pressure in the Gulf of Bothnia ("hungry horse"). Bottom: A gash as a result of collision with an iceberg (Reduta Ordona, 1996).

Damage statistics

Based on 486 cases of ice damage registered in DNV's database the various forms of damage are distributed such as shown in Figure 6.30.

As another example of typical damage can be mentioned the occurrences in the Gulf og Bothnia in 2003. That year, 24 damaged ships were registered, with the damage distributed as follows:
- 7 cases of damage to propellers
- 8 collisions between ships
- 3 cases of hull damage
- 4 cases of damage to engine / steering engine
- 2 cases of damage as a result of towing

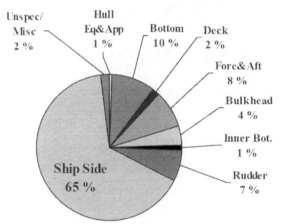

Figure 6.30
Distribution of ice damage registered by DNV. Total cases of damage are 486.

From Figure 6.30 we can see that the most usual damage occurs to the hull outside the bow region. Ice pressure along the sides is a usual cause, but during sailing in convoy it often occurs that ice floes float up under the bottom of the ship and cause damage.

6.1.5 Estimation of safe speed.

On the basis of experiences and calculation software, hull deformation can be presented as a function of the speed against a given ice mass. By analyzing the curve which is

shown in Figure 6.31, the speed that is required to perforate the hull will be ascertained. Consequently, safe speed can be regarded as somewhat lower than that. However, I would point out that such presentations are fraught with weaknesses, and cannot replace experience. However, the calculations can be a good supplement in an overall evaluation of the determination of safe speed. In order to give a picture of the size of the ice mass referred to in Figure 6.30, the starting point can be an ice floe which is 10 m in diameter, and 1.2 m thick – the mass will then be approximately 85 tons – a relatively small ice floe.

The hull load will also be able to be monitored electronically with the aid of sensors in the bow region and ice belt. Such installations have not gained ground in practical use yet.

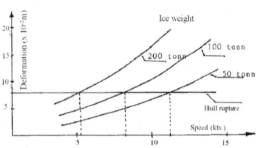

Figure 6.31
The hull deformation as a function of the ship's speed and the ice mass. The cracking point of the hull (horizontal line) is known and one can estimate safe speed (Kølher, 1986).

"Ice Passport"

The concept of the *Ice Passport* has been used by the Russians for many years. This defines the safe speed to a given merchant ship along a given route in ice-covered waters. The background for this is again to be able to use a ship carrying an ice class in relation to what is strictly worse than what the class dimensions for, provided that there are other restrictions, such as lower speed. The Ice Passport is therefore not a construction standard, but a description of how the ship can be operated. Russia will normally require such Ice Passport for all ships visiting Russian waters in winter, as well as all those that sail along the Northern Sea Route. This is provided that the ice conditions are more difficult than the criteria for the given ice class of the ship. The Ice Passport will contain a description of operating in convoy, including maximum allowed speed in various ice conditions (Figure 6.32), distance to the icebreaker, etc. The maximum allowed speed will not exceed what is regarded as safe speed or attainable speed. It will be the CNIIMF (Central Marine Research and Design Institute) in St. Petersburg that will issue the Ice Passport on behalf of the Russian authorities.

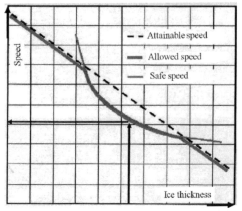

Figure 6.32
The diagram illustrates maximum allowed speed under various ice conditions, such as is the philosophy in an Ice Passport (DNV).

6.1.6 Ice anchoring and supplementing of water supplies

Anchoring in waters where there can be a danger of being frozen in or touching larger ice masses will often be an unwise act. There will be a great danger of imposing extreme loads on the ship. One alternative, however, can be to moor / anchor to a larger ice floe. This can be done by pulling

relatively small ice anchors up on the ice and mooring in the lee of the floe. When anchoring in this way it is of course essential to keep a close eye on other ice masses and the depth conditions which can be dangerous for the ship. It can also be possible to moor to provisional bollards of logs that have been frozen fast in holes in the ice (Figure 6.33).

Figure 6.33
KV Svalbard moored to a large ice floe where logs are frozen fast in the ice.

In the summer, large pools often form on top of the old drift ice (Figure 6.34). At the foot of pressure ridges this can be an excellent opportunity to supplement the ship's water supply. If ice has to be melted to obtain fresh water, it is important to see that it is old, and thereby almost free of salt.

Figure 6.34
Ponds on the top of old ice will be fresh and can be drunk. Here from the North Pole there is frozen thin ice on top.

6.1.7 Dynamic Positioning (DP) operation in ice

Ordinary DP systems will normally build up a model of environmental forces which affect the ship and regard this as current forces. However, this process will have great problems with handling larger loads from drift ice. Consequently, one cannot just operate ships with DP in ice. Several attempts have been made at this and operations with "ice management" has proved to function when the ice is not too difficult. Ice management means that an icebreaker breaks up / crushes the ice before it drifts towards the DP vessel (Figure 6.35). In that way the load from the ice will be moderate and have a relatively even course. Further, it will be important that the ship's heading is towards the direction of ice drift, and that the performance of the DP system is monitored closely. In that connection it can be relevant to disengage the degree of freedom in the "surge" direction so that with the joystick one can meet the ice forces with quicker and more powerful responses according to need. The method of ice management and sporadic manual joystick steering was used with a positive result when the *Vidar Viking* and the *Oden* carried out a drilling operation in NE-Greenland in the summer of 2008 (Figure 6.35).

Figure 6.35
Vidar Viking on DP in the ice at Northeast Greenland and Oden performs ice management (Photo: Viking Supply AS)

6.2 *Icing and stability*

Icing can create both operational and technical problems for ships sailing in arctic regions. There follows a description of the icing process itself, as well as the operational problems which can arise.

6.2.1 Icing

Icing can be caused by atmospheric water particles or sea spray. Atmospheric icing which arises because of supercooled fog or rain is seldom of any significance for the ship's stability, but can create problems for the performance of antennas for navigation and communication etc. (Figure 6.36).

Figure 6.36
Icing on antennas can be caused by atmospheric icing, sea spray or precipitation. The result can be reduced / zero performance of affected systems.

On the other hand, sea spray is a great threat to a ship's seaworthiness in arctic regions. Icing because of sea spray occurs particularly in strong winds and low temperatures, however, not when the temperature falls lower than 17 - 20 degrees of frost. At such low temperatures the water drops will freeze in the air and not be capable of attaching to rigging and superstructure. In heavy seas where there is more than normal spray, icing will occur even at very low temperatures.

In addition, the degree of icing is determined by meteorological and oceanographical conditions, movement in relation to the weather and the vessel's construction. There are several diagrams prepared for operational analyses (e.g. Mertin's diagram) for the prediction of the degree of icing. The diagrams in Figure 6.37 are based on ice reports from relatively slow-moving trawlers in the Barents Sea.

From Figure 6.37 the degree of icing can be predicted based on wind strength, sea temperature and air temperature. The degree of icing is then defined as follows:
1 - no icing
2 – little   1 – 3 cm/24 hours
3 – moderate   4 – 7 cm/24 hours
4 – extensive   7 - 14 cm/24 hours
5 – very extensive   15 – 24 cm/24 hours

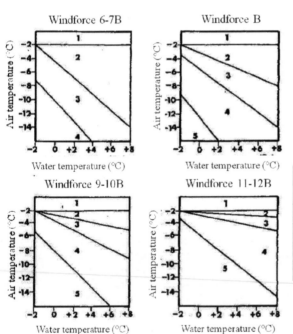

Windforce 6-7B   Windforce B

Windforce 9-10B   Windforce 11-12B

Figure 6.37
Prediction of the degree of icing (Mertins Diagram).

Example: Forecast – wind strength 9-10B, air temperature -8°C, water temperature +3°C
Expected icing – extensive, approximately 7 - 14 cm/24 hours.

There are also empirical formulas to be found which are used in some environments to predict the degree of icing, PPR. This is shown as:

$$PPR = \frac{V_a(T_f - T_a)}{1 + 0.3(T_w - T_f)}$$

where:

PPR   = Icing parameters
V_a = Wind strength (m/s)
T_f = Sea water freezing point (-1.8°C)
T_a = Air temperature (°C)
T_w = Sea water temperature (°C)

The connection between PPR and the rate of icing as previously presented will then be as shown in Table 6.1.

Table 6.1
Connection between the icing class, icing rate and PPR.

Class	Rate (cm/hr)	PPR
No	0	< 0
Little	< 0.7	0 – 22.4
Moderate	0.7 – 2.0	22.4 – 53.3
Extensive	2.0 – 4.0	53.3 – 83.0
Extreme	> 4.0	> 83.0

As the operator of the vessel there is little one can do with the environmental parameters in the diagrams. However, there is a lot that can be done with the degree of icing. First and foremost, one must reduce the danger of sea spray. This can be done by reducing the speed and keeping a course where one avoids waves hitting from between 15 and 50 degrees on the bow such as is shown in Figure 6.38. In addition, it is of great importance to have as much freeboard as possible. Consequently, larger merchant ships will be less exposed to stability-critical icing than that shown in Figure 6.37. On larger ships, however, large local loads will occur as a result of ice, as well as that rescue and safety equipment can be completely blocked (Figure 6.39). In such cases, the ship, per definition, is not seaworthy. On large ships, great distances to the icing location could also lead to under-evaluation of the problem (Figure 6.40).

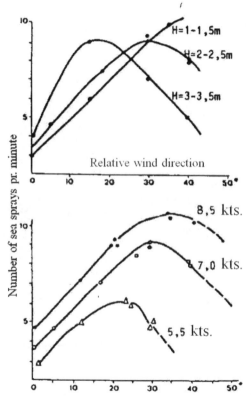

Figure 6.38
Significance of wave height (H) and speed
for the sea spray frequency as a function of
relative course.

Figure 6.39
Example of icing which has blocked the
lifeboat. In such cases the ship, per
definition, is not seaworthy, and attempts
must be made to remove the ice (Source:
Norwegian Coastguard).

Figure 6.40
The view from the bridge of this tanker
looks undramatic. However, at ground
level the situation is clearly more serious
(Source: DNV).

In ice-covered regions, icing on rigging and superstructure will seldom be a problem. This is because of reduced speed and little or no wave height. On ships equipped with a bubble system along the sides of the ship, a layer of ice can form from the waterline and a little way up the ship sides, but this will not have any effect on the stability.

In the case of normal icing the centre of gravity (G) will always be raised. Most often, the centre of gravity will also be moved horizontally (both fore and aft and abeam). These two factors will together lead to a considerable amputation of the GZ curve (the curve of the ship's righting arm). This is shown in Figure 6.41.

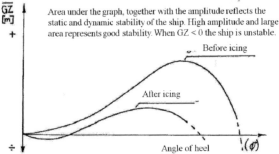

Area under the graph, together with the amplitude reflects the static and dynamic stability of the ship. High amplitude and large area represents good stability. When GZ < 0 the ship is unstable.

Figure 6.41
GZ curve, before and after icing.

Example:
The expeditionary vessel *Polar Circle* is loaded to standard condition as specified in Figure 6.42. The ship has containers on deck and 3 slack tanks. The stability in this condition is calculated in the ship's stability handbook from which the information is obtained. We take the case that the ship is exposed to heavy icing and 140 tons of ice has accumulated on deck, containers, rigging and superstructure (15.7 m above the keel). The icing will result in the centre of gravity (G) being raised and the GM is reduced. The link KG*sinα which is recognized from marine engineering will therefore increase and GZ will be reduced. The centre of gravity in this case will be raised 25 cm, which will mean a modified

GZ curve (broken line). We see that icing alone will not lead to any critical situation. However, the danger often arises when several unfortunate events take place at the same time, for example, hull damage which entails a completely different stability characteristic (leak stability). Use of anti-rolling tanks will also reduce stability. The tank on this ship takes 88 tons (operative) which entails a "free liquid surface moment" of 1439 tm. This relevant loading condition will entail a reduction of the metacentric height (GM) of approximately 32 cm. Complete figure examples for calculations will be found in the ship's stability handbooks. The calculations are performed according to the following formulas:

$$GG_2 = \frac{M_2}{\Delta}$$

$$GG_1 = \frac{W \bullet a}{\Delta + W}$$

GG_1 = centre of gravity (G)raise due to icing
GG_2 = apparent raise of G because of liquid surface in anti-rolling tank
W = weight of the ice
a = distance from the G of the ice to the centre of gravity of the ship
Δ = displacement of the ship
M_2 = the moment of the free liquid surface in the tank

On some ships (including the Viking ships) examples have been prepared which are presented as placards on the bridge and which can quickly be used as decision support in critical cases.

Other class requirements
The classification companies have focused mainly on the ship's structural ability to resist ice. Recently, however, there has been greater awareness of other conditions regarding operations in cold regions. DNV has therefore introduced some new class notations:

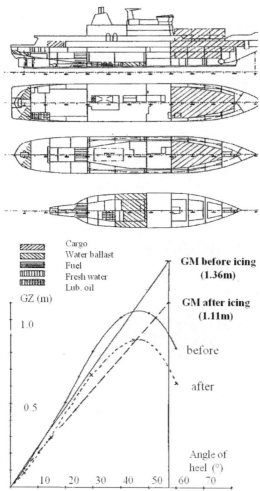

Figure 6.42
Advance calculated stability example and GZ curve with icing. From the ship's mandatory stability handbook.

WINTERIZED (-XX) and WINTERIZED ARCTIC (- XX). These are notations which entail that the ship shall be able to operate safely and without problems in daily mean temperatures down to a specified limit (-XX). In this is also normally included the notation DAT (-XX) which indicates Design Ambient Temperature and DEICE which

indicates that the ship has a built-in de-icing system which shall keep the safety-critical equipment free of ice. Such installations will normally be electrical heating cables or pipes circulating heated water. In addition, one will normally find the notations:

- OPP-F (Oil Pollution Prevention – Fuel).
- CLEAN which provides a standard for limitation of emission to the surroundings.
- RPS (Redundant Propulsion Separate) which describes a double and independent set of propulsion and steering systems (rudder, propellers, engine room, etc.).

In addition to this, on ships that operate in ice it should be evaluated whether one should not strive to get a construction which reduces the, at times, enormous vibration and noise which will be in a ship sailing in ice. DNV has a class for this which has the improved comfort for people on board especially in view. This is the notation COMF (Comfort).

What in practice is regarded as"winterizing" will be measures to keep the ship and equipment free of ice. In this is included rescue equipment, winches, firelines and tank ventilation. The measures can be:

- Covering with tarpaulins or placing in closed rooms.
- Draining of the rudder system and ventilation.
- Heating/ De-Ice with the aid of electrical heating cables or steam pipes.
- Cold-resistant oils or similar.

An example of ships where this has been paid a lot of attention is the supply ship *Viking Avant* (Figure 6.43) which was built for Eidesvik (www.eidesvik.no) for operation in the Barents Sea and similar places where icing must be expected. The ship carries class notation in DNV:

+1A1 Ice C SF Oil rec LFL*COMF-V(3)
E0 Dynpos-AUTRClean DK(+) HL(2.5)
DEICE-CStandby Container.

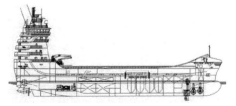

Figure 6.43
The Viking Avant is specially built with
regard to operations in cold climates. On
the photos we see the tarpaulin which can
be laid over the loading deck.

Operational draught
An ice class will be valid only if the ship
operates with a draught which is in
accordance with the ice reinforcement. If ice
forms on the side of the ship over or under
the reinforced area, this could entail damage.
The most obvious is that the draught must
not be so shallow that the propeller and
rudder can be exposed to heavy ice. One can
often wish to have the greatest possible
draught since this will give better propulsion
by the propeller and thereby improve the
ability to navigate. However, the draught
must not exceed the hull reinforcement.
Maximum draught is usually marked with a
symbol on the side of the ship (Figure 6.44)

Figure 6.44
Marking on the ship side indicates
maximum draught when operating in ice.
The example is from the Vidar Viking.

6.2.2 Stability when icebreaking

Icebreaking or entering into heavy pack ice
will mean a new stability condition which
physically resembles grounding. The
problem is visualized in Figure 6.45. There
can clearly be seen that a load is imposed
which in the centre of gravity measurement
can be regarded as negative. Consequently
there will be a raise in the centre of gravity,
with a resultant reduction of the GZ arm. If
this is in addition to an already amputated
GZ curve because of icing, the condition
will easily become critical. It is usual that
the icebreakers have prepared KG limit

curves (Figure 6.46), where the significance of lifting forces is included in the calculation. In the polar pack ice, this will always be the relevant condition since one will often have to sail the ship through larger or smaller pressure ridges. When the ship is stationary after entering the ice, one can expect considerable heeling until the ship has slid back or sunk through the ice.

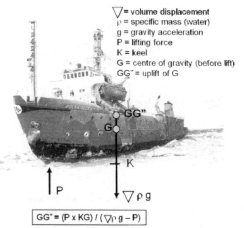

$\bigtriangledown$ = volume displacement
ρ = specific mass (water)
g = gravity acceleration
P = lifting force
K = keel
G = centre of gravity (before lift)
GG" = uplift of G

$$GG" = (P \times KG) / (\bigtriangledown \rho\, g - P)$$

Figure 6.45
Lifting force under icebreaking reduces stability.

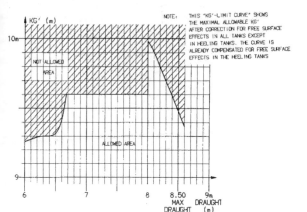

NOTE: THIS "KG'-LIMIT CURVE" SHOWS THE MAXIMAL ALLOWABLE KG' AFTER CORRECTION FOR FREE SURFACE EFFECTS IN ALL TANKS EXCEPT IN HEELING TANKS. THE CURVE IS ALREADY COMPENSATED FOR FREE SURFACE EFFECTS IN THE HEELING TANKS

Figure 6.46
KG limit curve from the stability handbook of the icebreaker "Oden". The effect of lifting forces when icebreaking is included in the calculation.

As a consequence of a hard strike with the arctic ice the danger of hull perforation will be present, if the ship is not constructed for icebreaking and ramming. Water penetration will consequently have a very unfavourable effect on the ship's stability. In literature, this is called leak or damage stability. Without going further into requirements and theory about this, it must be emphasized that such possible water penetration is probably a possible cause of the stability reduction of ships that sail in ice-covered waters.

Arctic shipping is especially exposed to reduction in stability. The situation must be continuously monitored so that measures can be taken at an early stage to keep the ship a seaworthy condition. Such measures can be removal of ice from rigging and superstructure. This can be done manually, by hacking (Figure 6.47), or with the aid of a de-icing system. Figure 6.48 shows a shocking example of unwarrantable icing.

Problems connected to leak stability can be prevented by keeping a safe speed. Double hull and several watertight compartments will reduce the effect of damage stability.

Figure 6.47
On smaller vessels it is usual to hack the ice off with a special ice pick.

Figure 6.48
Heavy icing on the forward part of a
merchant ship while sailing in the northern
Pacific (Source: Hoegh).

6.3 Icebreaking and convoy operations

In order to maintain rational shipping in ice-covered regions, it is often necessary to organize the ships in convoys. The convoy can be led by one or two icebreakers. Experience shows that convoys of any more than 5 ships are not practical. If the convoy meets difficult ice conditions, the ships at the rear can easily be trapped by the channel closing in. In the case of difficult ice conditions in Siberia, a nuclear icebreaker is often used first, then a diesel icebreaker before the convoy itself (Figure 6.49).

Figure 6.49
A convoy in difficult ice on the Northern
Sea Route. A nuclear icebreaker goes first,
then comes a powerful diesel icebreaker
right in front of the merchant ship.

In relatively short passages it can be necessary to keep a channel open with regular "ploughing" (Figure 6.50). This is a method which is very suitable in homogenous first-year ice, and is often used in harbour areas and on shorter routes in sub-arctic regions. In the Arctic, however, the ice is often moving so that "ploughing" in that sense is not relevant.

Figure 6.50
"Ploughing" of channels. The photo shows
the Russian icebreaker Madyug testing
performance at Svalbard (Source:
Thyssen/Waas).

For ships that sail in convoy, strict requirements will be made regarding operational preparedness. In practice this will often mean that on the bridge there must be two officers plus a lookout on watch. This standard is used by most operators in ice-covered waters. Since the margins in convoy sailing are small, one must always have an officer on stand-by at the engine /propeller control to avoid chain collisions (Figure 6.51). The other will be able to operate communications, navigation and searchlights. A good bridge design will be of extremely practical and safety-related significance in convoy operations.

The channel behind the icebreaker will always be more or less filled with ice floes. It is also usual that the channel closes because the ice masses float together again after the icebreaker. For ships that lie far behind in the convoy this will lead to a side pressure on the sides of the ship. When this happens, it will require a lot of engine power to sail further in the channel. The result will very often be that one can be caught in the channel (Figure 6.52). In a channel which is used a lot by icebreakers and convoys, finely crushed ice will accumulate. After a long time this could be thick and impenetrable "slush", especially in the vicinity of harbours and other junctions (Figure 6.53).

Figure 6.51
Collisions between ships can easily if the convoy stops. Here a collision between Yevgeni Titov and Bremer Saturn in 2003.

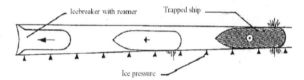

Figure 6.52
Ships can be caught in a channel by the side pressure from the ice.

Figure 6.53
After a lot of use, crushed ice will be accumulated in the channel.

In principle there are four methods to reduce the danger of being caught in a channel:
1)
First and foremost the distance between the ships can be reduced. The danger of collision, however, will be greater.
2)
In difficult conditions it is also usual to link the ships together as shown in Figure 6.54 (close couple towing). This can be done by attaching a cock's foot to the following ship. In order to get a favourable angle on the cock's foot the most suitable way may be to thread the towline up through both hawse pipes and fix it to the forecastle head. The anchors are then lowered so that the stock is clear of the hawse pipe. Such operations will impose great local loads on the ship.

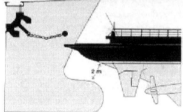

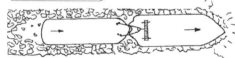

Figure 6.54
In difficult ice conditions it can often be most effective and safe to link the ships together. Above, from the bridge on the icebreaker Nordica during towing of ships.

In the propeller wake of the icebreaker, ice floes will be accelerated under the ship behind. Since the ice class does not make strict requirements regarding plate thickness on the bottom as for the ice belt, the ice floes will often make dents on the bottom of the ship. If the ship being towed is loaded, the icebreaker's towing notch will often come over the reinforced ice belt in the bow. This will impose such great loads on the ship that plates can be deformed.

When the ships are towed close to the stern, the ship in front will often have a side force imposed from the one at the back (which can also use its engine to push) and thereby gear the opposite way in relation to the one behind. If one is not vigilant, such a situation will easily develop into chaos. When the foremost ship (the icebreaker) discovers that it is about to lose control it can slacken off on the tow so that the ship at the end gets increased resistance, at the same time as the ship at the back reduces engine power. Communication between the ships in the convoy is therefore extremely important.

3)
Another method of avoiding being caught in the channel is to increase the speed of the convoy. However, this requires that the icebreaker has power reserves and that the ship can tolerate hitting smaller ice floes at greater speed. "Safe speed" must nevertheless not be exceeded. The danger of collision will also increase when the speed increases.

4)
A fourth method of improving ability to navigate is to improve the technology, i.e. by developing icebreakers that clear better channels (Figures 6.55 and 6.56), or developing merchant vessels that have less ice resistance.

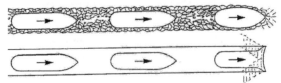

Figure 6.55
The above figure shows an attempt to illustrate the significance of the icebreaker's hull design to reduce the amount of ice in the channel.

Figure 6.56
Icebreakers with two azimuth propellers can throw the propeller water out a little, and thereby make the channel considerably wider. Here shown by the Finnish icebreaker Nordica.

A lot of research has been done in the field, and especially new bow designs and propeller systems will be of great significance.

In channels carrying a lot of traffic, after a time of icebreaking, there will more or less be "ice slush" in the channel. It can then be difficult to see where the last channel goes. It is therefore very important to be extra vigilant in the dark. Good searchlights will be absolutely necessary. If the ship leaves the channel in such ice slush, the merchant ship will surely get stuck.

If a ship does get caught in the channel, it will normally require icebreaker assistance to get free. If the convoy is escorted by only one icebreaker, this will affect the progress of the other vessels. One method which can be effective for freeing the stuck ship can be to manoeuvre the icebreaker as shown in Figure 6.57. The icebreaker then cuts a channel parallel to the original one. This will reduce the side pressure on the ship that is stuck, and it will hopefully be able to continue under its own power. As shown in Figure 6.56, icebreakers with azimuth propellers can use them actively in icebreaking tactics. This can be abeam flushing to open channels both in front and along the side of the ship that is stuck. If icebreakers with azimuth propellers get stuck, and must ram, experience has shown that ramming in heavy ice is often easier if the propeller water is thrown at an angle forwards along the sides of the ship before the ship backs out to undertake new ramming.

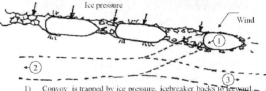

1) Convoy is trapped by ice pressure, icebreaker backs to leeward.
2) Icebreaker open lead to reduce ice pressure.
3) Icebreaker continue, and orders convoy to give full ahead.

Figure 6.57
Freeing of a ship wedged in the channel behind an icebreaker.

It will always be the icebreaker that leads the convoy. With its extensive experience and good access to information, the icebreaker will be the ship best qualified to plan the route. The information can be obtained from satellites, aircraft, other ships or the helicopter which one of the icebreakers often carries.

For more information on the icebreaker service in Baltic and Canadian waters, visit:
www.baltice.org
www.ccg-gcc.gc.ca

"Ice management"
In connection with fixed installations there can often be a need to clear the ice so that the load on the construction is reduced. Such operations are usually called *Ice*

management. How this is done will depend on the problem one is faced with. An example of this is towing of an iceberg that has been on a collision course with platforms (Fig. 6.58). Such towing is started by laying strong floating hawsers around the iceberg.

Another example can be to break up heavy and dense ice masses which drift towards a construction that is not dimensioned for such loads. This was done during the Arctic Coring Expedition in 2004. The relatively light icebreaker *Vidar Viking* was then equipped with drilling equipment to take core samples from the bottom of the Arctic Sea close to the North Pole. This ship, which does not carry an ice class for operating alone in such regions, had to lie in a constant position during the drilling operations. In order to make this possible in the drifting ice it was necessary that the ice which drifted towards the ship was broken up. First and foremost, the Russian nuclear icebreaker *Sovetskij Sojuz* was used for this work. The Swedish diesel icebreaker *Oden* did the job of further crushing of smaller and lighter ice. In this way the *Vidar Viking* was able to perform the work as planned (Figures 6.35 and 6.59). On *Vidar Viking* an attempt was made to use dynamic positioning (DP), but the effects of the heavy ice made this extremely difficult.

Figure 6.58
The Mærsk Gabarus tows an iceberg weighing approximately 5 million tons off Newfoundland in order to avoid collisions with platforms.

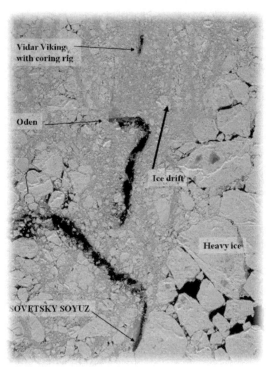

Figure 6.59
Aerial photo from Arctic Coring
Expedition where two icebreakers are used
t o break up the drifting ice to prevent large
ice loads on ships drilling in the seabed at
the North Pole.

6.4 Ice monitoring and route planning

Good knowledge of the ice situation is
essential in planning and performance of
marine operations in ice-covered regions. It
will also be an important element in more
complex ice-management systems.
Traditionally, on board the ships one could
receive an ice forecast in the form of radio
facsimile from ice-centres on shore (Figure
6.60). There have also been statistics
available from previous years, which have
included information regarding deviation
from the normal situation.

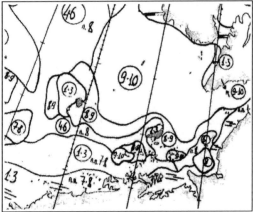

Figure 6.60
Handwritten facsimile of the ice situation
in the Northern Sea Route, September
1991. The chart is dispatched from the
Meteorological Centre in Dikson and must
be regarded as relatively low resolution
data.

It can be appropriate to divide ice
information into the following three
categories, based on degree of resolution
and accuracy:
- Statistical data.
- Low resolution data.
- High resolution data.

Statistical data. On the background of
statistics regarding ice conditions from
previous years, one can also make an
approximately estimate of the conditions
which can be expected. Such statistics can
be useful and essential for planning of
marine operations. They can be especially
useful in the marginal ice zone, where as a
rule one wants to perform an operation with
the least possible interference from ice. On
the statistical basis one can also calculate an
"ice window". The equivalent methodology
is used a lot during operations in the oil
industry where there is a necessity for a
"weather window" (ref. Kjerstad, 2010),
which is the probability of a limited wave
height in a given period. The ice window is
a calculation of the probability of not having
ice problems during an operation lasting for

a given period. The variables in the calculation of such ice window will be the length of the "open" period, acceptable limits for interference from ice, and acceptable probability. Figure 6.61 shows how such ice window in principle can be illustrated. Figure 6.62 shows another variant which can be used in more long-term planning. Here the variation in normal ice thickness over one year with several scenarios for climate development up to 2050 is shown. The horizontal lines indicate maximum navigability for different ice classes (PC3 – 5). The area which lies under the horizontal line can be regarded as a period of time in which the ship can operate in the "ice window". The figure is based on a study by DNV.

Another way to carry out strategic planning is to use an ice chart based on a normal situation, at the same time as one knows the standard deviation on this. If in addition there is information regarding other meteorological and oceanographic conditions, it will be possible to make long-term prognoses of the margin ice zone. The method described above will also easily be adapted to calculations of probability and prognoses for the possibility of colliding with an iceberg. If one is also to evaluate possible operations areas for various vessels, it can be useful to have access to ice charts which show the normal thickness and age of the ice. An example of a chart with an illustration of ice thickness is shown in Figure 6.63. The chart in question was used in a study performed by DNV in connection with the future potential for shipping over the Arctic Sea (Larsen, 2009).

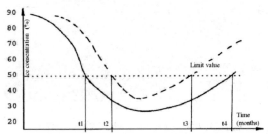

Figure 6.61
An example of how an "ice window" can be illustrated. During the period between t₁ and t₄ we have a given probability for acceptable conditions (under the limit) in a normal year, while in a difficult ice year we can expect acceptable conditions during the period between t₂ and t₃ (broken line).

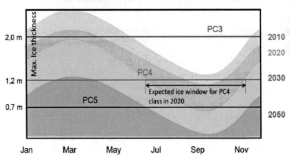

Figure 6.62
Possible development of an ice window for different ships, provided that there is an expected ice reduction (Source: DNV).

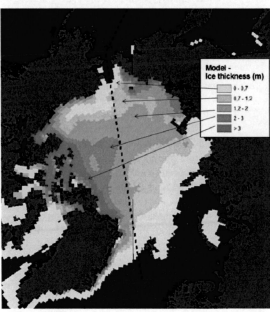

Figure 6.63
Chart of the average thickness of drift ice in summer (DNV).The broken line shows a possible route over the Arctic Sea.

Low resolution data. (Resolution of 5 - 50 km) When one has planned an operation from statistical data and has reached the phase of implementation, it is important to know where the ice edge is and how it is moving. In operations that are planned in the short term and in voyages in the marginal ice zone, the information regarding the ice edge can be sufficient to perform the voyage in an efficient manner. A typical example of low resolution data is the information from the NOAA satellites. Normal sensors can be optical sensors or scanning microwave radiometers. The limitation of the optical sensors is that they are dependent on clear weather. Information regarding the concentration of the drift ice in % of the total cover can also be analyzed from satellites with low resolution. Ships that find themselves in the area concerned can also give useful data in order to calibrate and verify data regarding the ice concentration. The ice chart shown in Figure 6.60 is also based on satellite data in this category.

The ice condition can change quickly under the influence of current and wind. On the basis of this, in 1985 an experiment was started for close real time transmission of satellite information to an icebreaker in the Gulf of Bothnia. One then read the satellite data from NOAA-6 and NOAA-9 at Tromsø Satellite station, transmitted data on a data line to a centre in Finland which transmitted the date further on the mobile phone network out to the icebreaker in the Gulf. The transmission appeared promising, but the limitations of the satellite system itself were too great to be of any particular use to the ice breaker in the daily route planning. In this is included especially bad resolution and dependence on clear weather. This type of date can nevertheless reveal great changes which take place over a relatively large area. An example of this is shown in Figure 6.64, where a 300 km long channel opened up in the area north of Svalbard in April 1999.

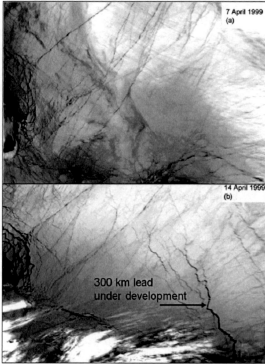

Figure 6.64
Optical satellite photos from the area north of Svalbard with 7 days' interval. A 300 km long channel which has opened up in the meantime can clearly be seen.

High resolution monitoring. On voyages in the Northwest, and especially in the Northeast Passage, one will be faced with a choice of route in different straits which are located relatively close to one another. In the one strait the ice can be tightly packed and in the other alternative there can be almost open waters. Such details will be difficult, if not impossible, to identify from data from passive satellites.

New perspectives for efficient route choices in narrow waters are opened with the use of the SAR satellites (Synthetic Aperture Radar). These are low-orbit satellites (approx. 550 – 800 km) which give far better resolution than ordinary meteorological satellites, and they will also be independent of the cloud cover. The resolution is calculated at ca. 30 metres in normal resolution mode. ESA operates three such satellites. These are ERS-1, ERS-2 and Envisat. In addition, data from Canadian and American SAR satellites will be available.

From SAR data one will to a certain degree be able to classify ice-age, -thickness, -movement, etc. This, together with the good resolution will help shipping to choose routes which minimize the ice load and increase safety. In many areas it is also reasonable to believe that there can be an improvement in the continuity of the traffic.

The limitation in the SAR data lies in the degree of coverage of the satellite. This can be seen in the track diagram shown in Figure 6.65. In addition, coverage is limited to a circle around the data acquisition station. Figure 6.66 shows that this will lead to an uncovered area on the east side of the Taymyr Peninsula in Siberia and West Greenland. However, coverage can be expected to be better since more stations will be able to acquire data in the future, such as there is now at Svalbard and in the Antarctic.

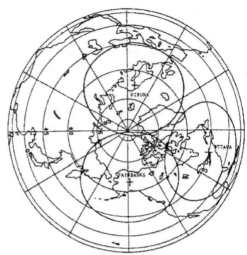

Figure 6.66
Possible coverage of ERS-1 data, acquired at Kiruna, Fairbanks and Ottawa. We see that coverage is lacking at West Greenland and the Taymyr Peninsula.

In order to illustrate the perspectives that lie in the use of SAR data as an aid to "ice routing" Figures 6.67 and 6.68 show that one can clearly see a pattern of channels in the Arctic pack ice. By comparing the photos drift vectors can also be generated as shown in Figure 6.69.

If one wishes to have even higher resolution than that described above, it can be achieved by airborne SAR or SLAR (side-looking airborne radar). The height of the aircraft will determine the width of the photo (track) and resolution. By flying the route in question one can quickly obtain data of high quality. The photos can be transmitted directly to the ships or icebreakers on a photo-transmission line. The method has been used extensively in a research connection, including in the Barents Sea and at Svalbard.

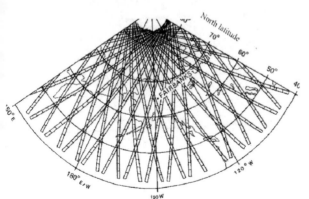

Figure 6.65
Projection of ERS-1, SAR coverage over a three-day period. Each track is 75 km wide.

How ice data shall be transmitted to ships in the Arctic is still an open question. With the technology that is found on ships at present (2011) it is most probable that data from the satellite is acquired by the shore station. Ready interpreted data will be adapted for the users there and transmitted as a compressed data file on satellite communication (Inmarsat, Iridium or similar). This is a relatively reasonable method for the user, but the resolution can often be considerably reduced in relation to the starting point. Satellite communication (Inmarsat) also has its limitations at high latitudes, as well as that the bandwidth means that very large amounts of data are not suitable for transmission. Many electronic chart systems (ECDIS) are now ready to present ice data as a separate information layer which can be shown together with the chart basis.

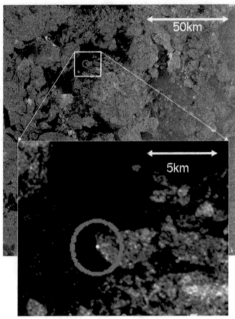

Figure 6.68
SAR data from the Fram Strait. On the enlarged photo below, the KV Svalbard can be seen as a white dot by a large ice floe (as in Figure 6.33). Source: KSAT.

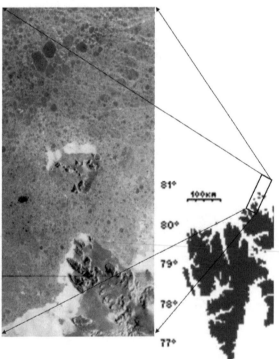

Figure 6.67
ERS-1 photo from the northern tip of Svalbard. The light areas are open waters.

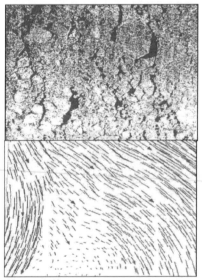

Figure 6.69
SAR data based on several pass is used to calculate the drift vectors of the ice (below) (Source: SeaSat).

152

If one has satellite photos available with a resolution such as shown in Figures 6.67 – 6.69 a trained eye will also be able to see most of the icebergs. This is because an iceberg will normally have another drift pattern than drift ice on the surface. Tracks or channels will therefore be made which often look like circles in the drift ice.

Data acquisition directly from the SAR satellite to the ship will mean large investments for the ship and at the same time will require qualified operators. Data acquisition to icebreakers and transmission to ships via radio-based photo transmission lines can be an alternative. With such a solution one will also be able to solve the problem of a gap in the coverage of parts of the Siberian coast and at West Greenland. The different ice data will in their separate ways also be an important element in a complete ice-routing system (Figure 6.70), dependent on which phase of the operation one is in.

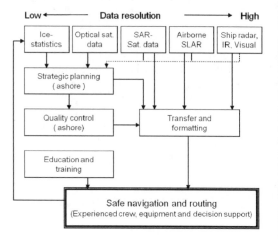

Figure 6.70
The diagram shows a block diagram of how a route planning system can be built up of different elements and at the end contribute to safe navigation.

Research is continuously being done to improve methods of collection of ice data and to better be able to classify the ice type.

For monitoring purposes, upward-directed sonars are placed on the seabed (ULS). An example of this is to be found in the Fram Strait along the 79th latitude, where one wishes to monitor the drift of the ice out of the Arctic Ocean. Another method which is continuously being perfected is the use of electromagnetic (EM) sensors for measurement of the ice thickness. These are sensors which can either be used directly on the ice or in an"EM-bird" under aircraft or helicopters (Figure 6.71). Such instruments have also been installed in front of the bow of ships in order to be able to compare the ice thickness with the loads which are imposed on the ship when sailing in ice.

Figure 6.71
Electromagnetic instrument for use in measuring the ice thickness – directly on the ice (below) or under a helicopter (above).

6.5 Search and rescue in the ice.

The alarm and search process itself in the case of casualties in the Arctic will in principle be no different from other areas in relation to what is described in the GMDSS (Global Maritime Distress and Safety System) and the IAMSAR manual. The difference lies in that these areas are not easily accessible and with far less available resources that one would find in more densely trafficated areas in the world. Those needing assistance must therefore prepare for a considerably longer response time. This requires than one is equipped to tackle the cold over a longer period of time. For ships that operate frequently in Arctic regions, therefore, it can be important to give thorough consideration to what this requires of equipment, over and above the mandatory equipment for the ship. During the development of the Polar Code which was the basis for IMO's "Guidelines for ships operating in Arctic ice covered waters" a manual was prepared in Canada for survival in cold regions (Marine Survival Handbook for Cold Regions) (Figure 6.72). This book should be found, and studied, on all ships sailing in arctic regions. The book can be downloaded from: http://www.tc.gc.ca/MarineSafety

Figure 6.72
Literature on survival in cold regions should be found on board.

One must also be prepared for the fact that lifeboats can be difficult to use, and that freefall boats will be unusable. These are circumstances which should be considered by every ship in its emergency procedures, and therefore included in its evacuation drills (Figure 6.73). The ship's company should also possess competence that enables them to repair damage which can occur to the hull. Over the years this has many times been essential for vessels that have been engaged in sealing (Figure 6.74).

If one moves out on to the ice or shore, one must be prepared to come across polar bears. In many instances the bear is not afraid of people and has to be frightened away. The procedure will then normally be to fire warning shots with a signal pistol at the bear (Figure 6.75). If this does not help, one must be prepared to shoot the bear. It is therefore important that one always has armed and trained personnel in the vicinity when the ship is left for one reason or another.

> **Remember!**
> Poorly developed SAR preparedness in polar regions should entail extra planning and caution during voyages and marine operations in such areas. Operating two ships concurrently (i.e. cruise) improves safety considerably.

Figure 6.75
Meeting with a polar bear on the ice or on shore can be life-threatening and one must always have armed and trained personnel in the vicinity.

If the ship is on expeditions in the Arctic where going ashore can be done, it will be worth participating in a safety course for this. Such courses are e.g. arranged at the University of Svalbard (www.unis.no).

Figure 6.73
When the cruise ship Maksim Gorkiy collided with an ice floe in 1989, all the passengers and most of the crew were evacuated (Source: Norwegian Coastguard).

6.6 Training and insurance-related matters

Up until now, education within arctic shipping and technology has been a neglected chapter in most maritime education around the world. Many places the traditions pertaining to the seal hunting environment have been the tools of communication in this area. It is from this environment many of the officers on the Arctic / Antarctic research ships have been recruited. The hunting environment over the past few years has been forced to reduce its activity, which means that the cornerstone in training and communication of arctic sea knowledge will be in danger of lapsing many places. In order to improve the situation, several maritime colleges around the world now (2011) offer courses in ice navigation. According to the STCW convention, all navigators shall be given training in planning and navigation for all conditions, including in ice. However, this is a relatively weak formulation which means that the curriculum in practice has been very

Figure 6.74
The sealer Polarfangst from Tromsø received a gash at the waterline. The crew managed to heel the vessel so that the hole was raised above the waterline and temporary repairs could be carried out. (Photo: Paul Stark, captain).

limited. In 2009, IMO made a decision to intensify this by including extended ice navigation in Part B of the STCW Code, with the intention of transferring it to the mandatory Part A in the near future. Courses are therefore being prepared which shall be in accordance with guidelines that are drawn up by the IMO ("Guidelines for ships operating in Arctic ice covered waters"). These guidelines have their starting point in the work with the "Polar Code" and will be required in the area shown in Figure 6.76. Further, an extension of the area to include the Antarctic is expected later.

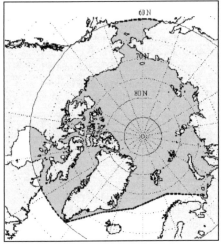

Figure 6.76
Area of applicability for IMO's guidelines for operation of ships in ice-covered regions.

The combination of continuous moving of the petroleum activity towards the north, and a more environmentally aware attitude in society will entail that in future there will be continuously more stringent requirements for the level of knowledge of operators of ships in these sensitive areas. Based on the perspectives from increased oil activity in the North, the classification societies now have drawn up their own guidelines for requirements regarding competence. In DNV this is a part of the SeaSkills Program. Education in ice navigation should also be a

matter of course within the shipping companies' ISM-certification when the ships are to operate in ice-covered and cold waters.

Arctic shipping is an area which makes strict requirements to knowledge, and it would be desirable to have a "management" notation to the ice class carried by a ship.

Large parts of the Arctic lie outside the area which is covered by the conditions laid down in the Insurance Plans. The ceding company, however, can grant a dispensation from this rule. If the policyholder does not accept the conditions, the insurance will not apply when the area of operation is exceeded. In paying compensation a deduction of a minimum of 25% will always be calculated. In addition, it is usual that the conditions of insurance contain formulations such as that sailing through ice must not take place except in channels that have been broken open for general shipping. Here it is worth noting that Russia regards the traffic along the Northern Sea Route as "general".

6.7 Some accidents

Over the years there have been a long series of accidents as a result of collisions with ice. Perhaps the most famous is the loss of the *Titanic* in 1912. Some of the incidents have already been described, but it can be useful to describe some in more detail. *Titanic* is not the only ship that has felt meeting the ice off Newfoundland. On the chart in Figure 6.77 there are plotted in 560 registered collisions with ice in these waters – several have led to sinking and serious damage.

KV *Andenes* on its mission in the Antarctic in 1989 – 90 (2 February 1990) the Norwegian coastguard vessel was trapped in drift ice in the Weddell Sea. The ice was in strong movement and pressed the 105 m long ship up from the ice in such a manner that for a time there was a danger that the ship could be lost (Figure 6.78). When the ice pressure eased off after a few hours, two of the ship's four propellers were destroyed, and the ship took in water. This was despite the fact that the ship was built to ice class 1A* in DNV. However, the crew was trained in repairing leaks and managed to stop the leak with cement. The ship could later return to Norway with half-speed.

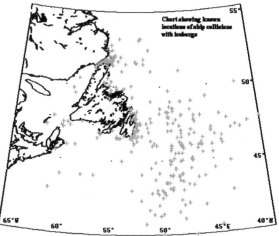

Figure 6.77
The crosses on the chart indicate positions where collisions with ice have been registered off Newfoundland.

Figure 6.78
On its mission to the Antarctic in 1989-90, the Norwegian coastguard vessel Andenes sustained serious damage as a result of ice pressure.

The cruise ship Hanseatic has run aground twice in the Arctic. First in Canada as a result of erroneous navigation, then caused by an unstable gyro at high latitudes. In July 1997 the ship ran aground again. This time outside a marked channel in Beinbukta in the Murchison fjord, northeast in Hinlopen at Svalbard (Figure 6.79). Running aground was a result of that the ship attempted to get closer to the wildlife there.

Figure 6.79
The cruise ship Hanseatic is pulled off byNorwegian coastguard vessel Tromsø after running aground at Hinlopen, by Svalbard.

On 19 June 1989**, the cruise ship *Maksim Gorkiy*** was on its way from Iceland to Magdalenafjorden at Svalbard with tourists. On the way there were open waters, good weather but bad visibility. At 2305 hours when the ship is sailing at almost full speed, at position N 77°37' - E 004°19' it struck a narrow drift ice belt, and the ship sustained a relatively large gash in the hull on the starboard bow. This was approximately 110 nm west of Isfjorden at Svalbard. The ship, which did not carry an ice class, shipped water, and it looked as though it was going to sink (Figure 6.80). Everyone except a core of crew members were evacuated to the lifeboats and put out on the ice floes around the ship. However, Norwegian coastguard vessel *Senja* was in the vicinity and pumps were taken on board, at the same time as divers stretched a tarpaulin over the damage to limit the leak. This allowed the ship to be assisted to Longyearbyen where better temporary repairs could be done. In all there were 953 people on board. If KV *Senja* had not been in the vicinity, and had there been a little more wind, the outcome could have been very dramatic. Luckily nobody was seriously injured.

Figure 6.80
The cruise ship Maksim Gorkiy with 953 people on board, came extremely close to being a total loss in 1989 because of striking ice. Below, evacuated passengers on the drift ice.

The KV *Senja* is on ordinary patrol at the Isfjord bank and the crew of 53 now have 2½ hours in which to prepare the ship for a situation which far exceeds the ship's capacity, which is normally handling of maximum 110 persons. Underway, KV *Senja* receives info from HQ in Bodø that an Orion aircraft and a Sea King helicopter will arrive at about the same time as the ship. Luckily, a Sea King has been stationed at Bjørnøya. Others are at Banak and at Bodø on mainland Norway. Weather: SSW-2, SS-1, stratus, moderate visibility, +1°C, swell 1-3 m. What they can encounter became clear to them underway. The heat on board is increased and bathtubs with hot water are made ready, hangars are prepared for reception and registration. Medicines, blankets etc. are prepared. The crews are briefed on first aid and how one can tackle a large number of wounded and dead.

To illustrate some of the challenges the KV *Senja* faced in the operation, there follows a summary of the log of the KV *Senja*:

0005 hours (20 June) – Transmission of distress call and request for assistance on 500kHz (MF emergency channel telegraphy). The watch receiver on the KV *Senja* had not been triggered. Why not a message on 2182kHz (telephony)?
0030 hours – KV *Senja* receives a MayDay call from C/S UYAD, pos. N77°30'-E004°09E from Svalbard Radio. Terminates fishing inspection and sets course at full speed (22 knots), is 77 nm away, ETA 0400 hours - Orion from Andenes and Sea King alarmed.
0111 hours – KV *Senja* receives a message from MG on weak VHF of hull damage and that the passengers are evacuated to the lifeboats.
0117 hours – Posisjon estimate adjusted to N77°37' – E004°19' (approx. 7.4 nm from the first position)
0130 hours - MG informs that she is sinking and has evacuated 953 persons (575+378) and that 5 lifeboats are on the water. Now the Coastguard becomes aware of the extent.

0318 hours – 325 passengers in rafts and lifeboats are observed between *Senja* and MG. Visibility is limited. It is the middle of the night and many of the passengers have only nightclothes and a jacket on. The lifeboats are in danger of being crushed under the constant movement of the ice. The crews work continuously to prevent this. The 5 rafts were taken on to the ice so that they would not be crushed by the ice.

0330 hours – the KV *Senja* enters the drift ice (multi-year floes) and reduces speed to 0 – 2 knots. Poor visibility because of fog. The ice belt is estimated to be 12 nm long and 1.5 – 3 nm wide. Concentration 8-9/10. Ice thickness 1 – 3m. The ice belt came as a surprise. There was no info from MG regarding ice.

0400 hours– MG is observed in open waters west of the ice belt – the bow is low in the water.

0420 hours – Rafts and lifeboats can be seen over a large area in the ice belt.

0428 hours – Distress flares are observed from an ice floe where there are 160 passengers.

0440 hours – The first lifeboat with approximately 120 passengers is alongside the ship and MOB boat is launched. Starting panic and difficult to get older persons on board. A line is shot over to the floe with the raft to tow it into the ship's side with the aid of the anchor winch. There is a light wind and 1-3m swell. Rescue net cannot be used.

Sea King and 2 Aeroflot helicopters have arrived and started rescue operations. The Orion has considered dropping rafts but climbs to 20 000 feet to direct air traffic by radar and be a relay station with land. The Orion's operations are regarded as very important for air safety.

0520 hours – Lifeboat No. 2 alongside.
0540 hours – MG requests diver assistance.
0550 hours – Lifeboat No. 3 alongside. Lifeboats are towed.

Sea King takes up total 50 persons from the ice and lands them on the helideck.

0600 hours – Fishing boat *Havfangst* arrives to assist.

0614 hours – Sea King lands on KV *Senja* for refueling and discharging of passengers. It is normally not legal to land this large helicopter on the vessel, but can be done because of the fine weather.

0625 hours – MOB boat encounters problems with reaching lifeboat No. 4 because of ice. Rescue man must run over a line. A passenger runs in panic towards the boat.

0635 hours – KV *Senja* has manoeuvred in to the ice to lifeboat No. 4 and has it alongside.

0655 hours – Sea King lands again to discharge passengers. .

0700 hours – Sea King checks that all rafts are empty, and then sinks them. MG requests bilge pumps. Lifeboat No. 5 has engine trouble because of ice.

0712 hours – Lifeboat No. 5 alongside. Bilge pump using 440V cannot be used on MG.

0730 hours – Super Puma with medical personnel from Tromsø arrives, refueling.

0736 – Sea King transports two diesel-powered bilge pumps and an engine assistant over to MG. Everyone is now picked up from the lifeboats. There are problems with towing the lifeboats in the ice. *Senja* leaves the ice and goes towards MG.

0740 hours – Refueling Sea King. There are now 3 Sea Kings, 1 Super Puma, 2 Aeroflot helicopters, one Orion and a Russian surveillance aircraft (May) in the air. Svalbard declares the area as limited.

0750 hours – MG wants to transfer the last of the passengers with the remaining lifeboat (still 5 on board).

0808 – Sea King takes over pump No. 2 and brings passengers back (some passengers will not let themselves be lifted up), refueling.

0830 hours– Passenger falls into the sea from MG. Picked up with lowered lifeboat.

0840 hours – MOB boat with three divers assist MG.

0900 hours – Lifeboat No. 6 alongside.

0904 hours – Lifeboat No. 7 alongside. *Havfangst* and *Polarsyssel* are on st.by in the area

0915 hours – Super Puma refuels and takes 17 prioritized passengers and films to Longyearbyen

0920 hours – Divers report a 6 m long and 10-30 cm wide gash on the starboard side and two small holes in the bow. The engine assistant reports that water is flooding through elevator shafts and doors that are not watertight.

0930 hours – All passengers have left MG

0958 hours – Sea King refuels

1000 hours – The situation is "stable" – 573 persons taken on board KV *Senja*. Havfangst is sent out of the area.

1028 / 1034 hours – MV *Breisund* and MV *Oknin* report that they are available.

1100 hours – Orion reports that 3 Russian vessels are underway to the area.

1135 hours – Divers report more leaks than pump capacity. MG lies deeper with increasing list. *Polarsyssel* still on stby. And preparations are made to evacuate the remainder of the crew.

1140 hours – Orion returns to Andøya.

1152 hours – MG deeper in the water. One pump stopped. Captain on MG arranges an information meeting. The water is now reaching up to Deck 6 and it is beginning to flow through a broken porthole in the galley, but this is successfully sealed with a blind porthole.

1158 hours – The engine assistant is flown back to *Senja*.

1218 hours – Sea King refueling.

1225 hours – Russian tugboat with 3 pumps is within range of vision.

1236 hours – MG prepares evacuation.

1249 hours – Super Puma arrives with two new pumps and three doctors. Returns with 15 passengers to Longyearbyen.

1311 hours – Sea King lands with 100 loaves of bread. Refuels and takes 14 passengers to Longyearbyen. The water is now at Deck 6 on MG.

1410 hours – Aeroflot helicopter arrives and drops sealing materials.

1415 hours – Captain on MG has agreed to evacuate more persons after repeated requests. Boat with crew from MG alongside. There are still 185 souls on board MG.

1429 hours – Observed that a lifeboat is filled, and then emptied again on MG – a lot of gesticulation and tense atmosphere.

1445 hours – Assists MG in getting the tarpaulin in place on the ship's starboard side.

1450 hours – Captain on MG reports stabilizing of water level.

1504 hours – It can be observed that MG is lighter at the bow.

1527 hours – Civilian aircraft arrive (Trønderfly), probably to film for the media.

1530 hours – Sea King refuels and takes 14 passenger to Longyearbyen (one on a stretcher).

1600 hours – Lifeboat alongside with several crew members of MG.

1700 hours – The Russian tugboat *Petsjenga* arrives in the area and salvages the three lifeboats which had broken loose in the ice.

1730 hours – Two Aeroflot helicopters arrive with more sealing materials / cement for MG.

1738 hours – Divers return to KV *Senja*

1743 hours – KV *Senja* leaves the area with 500 German passengers (73 sent med helicopter) and 193 crew members from MG. The trip is done at full speed to Longyearbyen where the Germans shall be transported directly home. There are still 185 crew members on board MG.

2355 hours – Svalbard Radio cancels the MayDay situation.

0120 hours (21 June) – KV *Senja* arrives at Longyearbyen.

0230 hours – Presse conference. 60 journalists have attended. Travel Guide Prinz rejects speculation of use of alcohol on board MG. He himself was on the bridge when the collision occurred.

0300 hours – The Governor arranges passport control, counting and transport to the airport. All passengers are now ashore.

Upon notification that the crew members shall be flown to Murmansk there are strong protects. After a meeting with the Consul and Governor, it was decided that KV *Senja* should transport the crew to Barentsburg so that they could go on board the MG later. KV *Senja* received military clearance from Norway to sail to Barentsburg.

0600 hours – Departure from Longyearbyen
0800 hours – Arrival Barentsburg. Crew members from the MG, as well as lifejackets etc. are put ashore.
0900 hours – Departure Barentsburg
1100 hours – Anchored in the roadstead at Longyearbyen. Crew are exhausted.
1800 hours – Press conference in Longyearbyen. Extreme media attention all day. Captain Sigurd Kleiven asserts afterwards to have had great benefit of the Info Officer from Defence.

Comments and experience from the KV *Senja*.

- Communication was challenging. HKO had to use 4 different radio networks because of different standards + internal communication. Aircraft radar faced challenges with ice clutter, especially with Russian helicopters without transponders.
- 15 landings were done by the Sea King and 3 by the Super Puma. Disadvantage that the ship's own helicopter and important crew for it were not on board. Almost 9000 litres of helicopter fuel were delivered.
- Doctors brought on board were a great help. In the beginning most were calm, but stress reactions set in after a time. Many were also exhausted (including after a pleasant evening cruise on the MG). Many elderly people had not been able to bring their important medicines with them. Approximately 170 people were treated in one way or another on KV *Senja*. None were regarded as critically injured. Stress reactions diminished when the situation stabilized,

but increased against as a result of the wait for the last clarification of the situation on MG (from approx. 1000 – 1800 hours). Language problems were challenging, but two Russian interpreters were very useful. Upon arrival at Longyearbyen the water supply was exhausted (even after full production), the toilets were blocked because of wrong use and overload.

- All communication with land had gone via Svalbard radio. Military radio traffic was dropped, partly because of bad conditions on the frequencies used. German travel guides who came on board after a time were largely used on the bridge to inform passengers over the intercom.
- Getting the evacuated persons on board was demanding. Only a few managed to climb in a net or ladder. The gangway was unusable because of rolling of 10-15°. With the aid of 4 men (2 in the lifeboat and two on deck) the persons were lifted in to the aft deck through a gate when the *Senja* lurched over. It took approximately 20 minutes to get 100 people on board. In total approximately 70 tons were lifted in this manner!
- Communication with the Captain of MG was difficult but became easier when he was assured that no costs would be involved for the rescue.

Food for thought
What would have happened if the KV *Senja* had been on patrol on the Norwegian coast or having a change of crew at main land? The distance is then approximately 560 nm – that means approximately 28 hours at 20 knots. It would also have entailed large limitations for the helicopters. The ship could have reached the casualty as 0430 hours on 21 June (+ any delays because of ice).
Without help from the Coastguard, the *Maksim Gorkiy* would probably have sunk at approximately 1200 hours on 20 June.

What if this had happened in another place (i.e. north of Svalbard or in between Kvitøya and Franz Josefs land)? What if the weather had been a little worse? The outcome would have been much more dramatic!

In August 2007, **the cruise ship *Alexey Maryshev*** suffered a serious accident on one of its weekly expedition trips to Svalbard. The ship was on a cruise at Hornbreen at Svalbard when the glacier suddenly calved (Figure 6.81). Parts of the glacier fell on the foredeck and 20 people were injured, some seriously. The accident shows all too clearly the dangers of sailing near glaciers and icebergs. In such situations there are small margins which separate the accident from total loss.

Figure 6.81
Passengers on the cruise ship Alexey Maryshev were seriously injured as a result of ice falling on the foredeck from Hornbreen at Svalbard. The photo is taken from the wheelhouse roof as the ship approaches the glacier.

In November 2007, **the cruise ship *Explorer*** was on a voyage in the Antarctic. At the Antarctic Peninsula (62°S) the ship, which was built to ice class 1A in DNV, entered a belt of drift ice and growlers.

This took place just after midnight and there was then no daylight. Striking the ice resulted in a gash in the hull so that large amounts of water flowed in. The ship heeled over quickly, but the 154 passengers got into the lifeboats before the ship went down (Figure 6.82). Luckily, the Norwegian coastal express steamer *Nordnorge* was only six-seven hours away and could come to the assistance of and pick up the hypothermic passengers. If the weather had been worse and other ships a little further away, the accident could have been a formidable catastrophe. The *Nordnorge* also assisted in saving another coastal express steamer, the *Nordkapp*, in the same area the year before. The *Nordkapp* had then run aground in an area not covered in maritime charts (ref. Chap. 3.1).

Figure 6.82
The cruise ship Explorer just before it sinks after having sustained ice damage in the Antarctic in 2007.

After the accident, the report by the flag state (Liberia) criticized the captain's evaluation of the ice conditions. He should have understood the large mixing of multi-year ice (glacial ice) and the dangers that this entailed. It was also emphasized that the ship should not have sailed in the dark under such conditions. The extensive efforts that were made to get the lifeboats out under extremely difficult conditions were also stressed as positive in the report. The report further emphasizes that Liberia will use its influence in the IMO to have stricter rules imposed as regards competence for voyages in ice-covered regions.

> The accidents with *Explorer* and *Nordkapp* in the Antarctic show all too clearly the significance of operating together with other ships.This ensures assistance in areas where other infrastructures are totally insufficientl.

6.8 Further development

Shipping and oil production are moving steadily northwards because of the lack, and less spreading, of ice. In Canada and Alaska oil production has been carried out in areas which parts of the year are totally covered in ice. In the Barents Sea and on Svalbard exploration drilling is carried out, and there is a lot of research being done on meeting the challenges which the tough conditions present. Concerning oil and gas production, it is nevertheless the development in Russia which will affect the activity here the most. Here there are large areas under development, and start-up both in Northwest Russia and at Sakhalin on the east coast. A little further into the future, when the arctic technology is better developed, there could also be talk about exploitation of known resources in more central parts of the continental shelf along the coast of Siberia. The fisheries have also adapted their technology to penetrate further in to the marginal ice zones, especially when locating shrimp and shellfish. We are also on the threshold of extensive exploitation of the krill occurrences in the Antarctic.
In 2007 we experienced the lowest ice spread that has been observed. In August, only 4.1 million square kilometers were covered by ice. Thereby the lowest record of 5.3 million square kilometers was thoroughly beaten. The trend was also similar in 2008, 2009 and 2010. If this development continues, one will be able to see completely new possibilities for shipping through the Northeast Passage and along the Northern Sea Route, as well

as that the cruise activity will have areas available where hardly anyone has been before. Routes in international waters across the Arctic Sea will also be investigated with new realism. Even though the sailing routes are short in relation to the Suez and Panama, one must nevertheless assume that the efficiency of relatively small ice-classed ships in many cases can have problems with the large specialized ships which sail at fast speeds on the southern routes (Figure 6.83). In the foreseeable future, therefore the development of regional shipping in the northern areas can have far greater significance than possible long transit voyages in a relatively short sailing season.

The development and possibilities for shipping in the northern areas will in the future have great geopolitical significance and create new trade-strategic alternatives. The greatest challenge in this development will nevertheless be to be able to operate in unison with the sensitive environment. This will require improvements in technology, readiness and the rules for operation.

Figure 6.83
It will be difficult for ice-classed ships (top) to be competitive in relation to the large specialized ships which sail through the Suez Canal (bottom).

6.9 Questions from Chapter 6

1)
Give an example of useful navigational equipment when sailing in ice.

2)
Describe operational conditions and preparations for sailing in ice and cold climates.

3)
What is meant by "ice blink"?

4)
When the ship is lying in relatively open drift ice, what can you look for to get a pre-warning of possible increasing ice pressure?

5)
What ice class must the ship carry in order to ram into dense drift ice?

6)
What must you think about especially when you are reversing a ship that has conventional propulsion?

7)
How would you adjust the radar in order to be able to observe small ice floes and growlers in otherwise open waters?

8)
What dangers are connected to radar positioning in areas with ice?

9)
What conditions would you emphasize in order to determine safe speed, and what is meant by an ice passport?

10)
How would you find information about the rules for convoy operations in the Gulf of Bothnia?

11)
What do you characterize as the most dangerous ice conditions for an ordinarily ice-classed merchant ship?

12)
What is the most usual ice damage on a ship?

13)
How can you ensure new water supplies on board when you operate in the Arctic Ocean in the summer?

14)
What is the condition for operating ships on DP in ice?

15)
What is meant by "ice-management"?

16)
What determines the degree of icing on a ship?

17)
How much ice can you expect to accumulate on the ship (typical ocean research vessel) when there is wind strength 8, air temperature of -8°C and the sea temperature is +2°C?

18)
How will the stability of a ship be affected when operating in ice-covered waters?

19)
What dangers are connected with sailing too close to glacier fronts and icebergs?

20)
What is contained in the class notation "winterized (-xx)"?

21)
There will be a triangular symbol with the inscription "Ice" on the side of the ship. What does this mean?

22)
How can ice that is frozen fast on the deck
and rigging be removed?

23)
What is the danger of sailing in an old
icebreaker channel in tracts near the coast?

24)
Give an example of how one can free the
ship after having got stuck fast in a channel
behind an icebreaker.

25)
What is meant by an"ice window"?

26)
What resolution (degree of detail) can be
achieved on modern radar-based satellite
photos?

27)
How can the thickness of ice be measured
from a helicopter?

28)
The ship has sustained a gash just under
the waterline as a result of a hard strike on
ice. What will you do to stabilize the
situation?

29)
What can be done to reduce the risk of
operating cruise ships in the Antarctic?

7 Reference and literature list

Alme, J., 2009. Ishavsfolk si erfaring: boka om is, isens menn, storm og forlis. (*Experiences of Arctic Ocean travellers: a book on ice, the men who work with it, storm and loss).* Tapir Akademisk Forlag, Trondheim

Armstrong, T., 1990. *The Northern Sea Route, 1989.* Scott-Polar Inst., Cambridge.

Brigham, L., 1990 : *The Soviet arctic marine*, Navale Inst. Press, USA.

Blindheim, J., Østerhus, S., 2005. *The Nordic Seas, Main Oceanographic Features. In: The Nordic Seas - An Integrated Perspective*. Washington, DC: American Geophysical Union.

CCG, 1992. *Ice Navigation in Canadian Waters,* Canadian Coast Guard, Ottawa.

Garrison, T. 2002. *Oceanography. An invitation to marine science*, fourth edition. Brooks/Cole, Pacific Grove, Ca.

Gudmestad, O.T., Løset, S., Alhimenko, A.I., Shkhinek, A., Tørum, A., Jensen, A., 2006. *Engineering aspects related to arctic offshore development*. LAN-design, St. Petersburg.

IDAP, 1988. Forskningsraport, *(Research Report).* Fjellanger Widerøe / NHL, Trondheim.

Herfjord, 1982. *Form design of vessels*, NIF-compendium, Norske Siv. Ing. forening, Oslo.

Ignatjev, 1966. *Screw propellers for ships navigating in ice.* Subostrojenie, St. Petersburg.

Kjerstad, N., 1990. *Drift av fartøy i Arktiske strøk, med spesiell vekt på skipsfarten i Nordøstpassasjen, (Operation of vessels in Arctic regions, with special emphasis on shipping in the Northeast Passage).* Tromsø Maritime College.

Kjerstad, N., 2006. *EGNOS - user experiences at high latitudes*. Article in the European Journal of Navigation, Vol. 4, Nr. 2.

Kjerstad, N., 2010. *Elektroniske og akustiske navigasjonssystemet. (The electronic and acoustic navigation system)* (4th ed.)Text book, Tapir Akademisk Forlag, Trondheim.

Kjerstad, N., 2010. *Fremføring av skip, med navigasjonskontroll. (Operation of ships with navigational control)* (2. ed.). Tapir Akademisk Forlag, Trondheim.

Knudsen, P., 2009. *Challenges in Geodesy / Geodynamics*. Lecture at the ESA meeting on GNSS at high latitudes, Tromsø.

Køhler, P. 1986. *Safe speed in the Arctic.* Veritec, Oslo.

Larsen, M.Mejlender, 2008. *Ice Load Monitoring*. DNV presentation on board the KV Svalbard.

Larsen, M. Mejlender, 2009. *Et åpent Polhav – en ny transportrute. (An open Arctic Ocean – a new transport route)*. Lecture at the Shipyard Conference in Ålesund.

Løset, S., 1989. *Oljens egenskaper.., (The properties of oil..)*. Vol.1, NHL, Trondheim.

Løset, S., Shkhinek, K.N., Gudmestad, O.T., Høyland, K.V., 2007: *Actions from ice on arctic offshore and coastal structures*. LAN-design, St. Petersburg.

Michel, B. 1978. *Ice-Mechanics*. Les presses de Lùniversite Laval, Quebec.

Nansen, F., 1924. *Blant sel og bjørn : min første Ishavs-ferd. (Among seal and bear: my first Arctic Ocean voyage)*. Dybwad Forlag, Kristiania (Oslo).

NP*). *Arctic Pilot* (NP-11, 23, 10). British Admiralty, London.

Peresypkin, V., Tsoy, L., 2006. *Latest development of design of icebreakers*. Presentation at Arctic Shipping – 2006, St. Petersburg.

Ugryumov, A., Korovin, V., Dale, S., 2005. *Tigu-su – På isflak mot Nordpolen. (Tigu-Su – On the ice floes towards the North Pole)*. Forlaget Nord, Tromsø.

Vargas, 1988. *Revolution in Polar shipping*. IAHR-proc., Sapporo.

Østreng, W., 1982. *Sovjet i nordlige farvann. (The Soviet in Northern waters)*. Gyldendal, Oslo.

*) Publications referred to as NP are the Nautical Publication series by the British Admiralty.

8 Index